Springer Theses

Recognizing Outstanding Ph.D. Research

For further volumes:
http://www.springer.com/series/8790

Aims and Scope

The series "Springer Theses" brings together a selection of the very best Ph.D. theses from around the world and across the physical sciences. Nominated and endorsed by two recognized specialists, each published volume has been selected for its scientific excellence and the high impact of its contents for the pertinent field of research. For greater accessibility to non-specialists, the published versions include an extended introduction, as well as a foreword by the student's supervisor explaining the special relevance of the work for the field. As a whole, the series will provide a valuable resource both for newcomers to the research fields described, and for other scientists seeking detailed background information on special questions. Finally, it provides an accredited documentation of the valuable contributions made by today's younger generation of scientists.

Theses are accepted into the series by invited nomination only and must fulfill all of the following criteria

- They must be written in good English.
- The topic should fall within the confines of Chemistry, Physics, Earth Sciences, Engineering and related interdisciplinary fields such as Materials, Nanoscience, Chemical Engineering, Complex Systems and Biophysics.
- The work reported in the thesis must represent a significant scientific advance.
- If the thesis includes previously published material, permission to reproduce this must be gained from the respective copyright holder.
- They must have been examined and passed during the 12 months prior to nomination.
- Each thesis should include a foreword by the supervisor outlining the significance of its content.
- The theses should have a clearly defined structure including an introduction accessible to scientists not expert in that particular field.

Karsten Brüning

In-situ Structure Characterization of Elastomers during Deformation and Fracture

Doctoral Thesis accepted by
Technische Universität Dresden, Germany

Author
Dr. Karsten Brüning
Leibniz-Institut für Polymerforschung Dresden
Department of Mechanics and Structure
Dresden
Germany

and

Institut für Werkstoffwissenschaft
Technische Universität Dresden
Dresden
Germany

Supervisors
Prof. Dr. Gert Heinrich
Dr. Konrad Schneider
Leibniz-Institut für Polymerforschung (IPF)
Dresden
Germany

ISSN 2190-5053 ISSN 2190-5061 (electronic)
ISBN 978-3-319-36034-8 ISBN 978-3-319-06907-4 (eBook)
DOI 10.1007/978-3-319-06907-4
Springer Cham Heidelberg New York Dordrecht London

Softcover reprint of the hardcover 1st edition 2014

Printed on acid-free paper

Springer is part of Springer Science+Business Media (www.springer.com)

Parts of this thesis have been published in the following resources:

K. Brüning, K. Schneider, S.V. Roth, G. Heinrich: Kinetics of strain-induced crystallization in natural rubber studied by WAXD: Dynamic and impact tensile experiments. Macromolecules 45 (2012), 7914–7919. Copyright © 2012 American Chemical Society

K. Brüning, K. Schneider, K. Heinrich: Deformation and orientation in filled rubbers on the nano- and microscale studied by X-ray scattering. Journal of Polymer Science Part B: Polymer Physics 50 (2012), 1728–1732. Copyright © 2012 John Wiley & Sons, Inc.

K. Brüning, K. Schneider, S.V. Roth, G. Heinrich: Strain-induced crystallization around a crack tip in natural rubber under dynamic load. Polymer 54 (2013), 6200–6205. Copyright © 2013 Elsevier B. V.

K. Brüning, K. Schneider, G. Heinrich: Short-time behavior of strain-induced crystallization in natural rubber. Proceedings of the VIII. European Conference on Constitutive Models for Rubber (ECCMR 8), 2013, 457–460, CRC Press, London. Copyright © 2013 Taylor & Francis Group, LLC

K. Brüning, K. Schneider: Effects of strain field and strain history on the natural rubber matrix. Proceedings of the VII. European Conference on Constitutive Models for Rubber (ECCMR 7), 2011, 29–32, CRC Press, London. Copyright © 2011 Taylor & Francis Group, LLC

K. Brüning, K. Schneider, G. Heinrich: In-situ structural characterization of rubber during deformation and fracture; in: M. Klüppel, W. Grellmann, G. Heinrich, K. Schneider, M. Kaliske, T. Vilgis (Editors): Fracture Mechanics and Statistical Mechanics of Reinforced Elastomeric Blends (Lecture Notes in Applied and Computational Mechanics), 43–80, Springer 2013. Copyright © 2013 Springer

Supervisors' Foreword

The description, and in particular the understanding of the highly nonlinear, and often anisotropic mechanical behavior of elastomers still presents a challenge to the rubber research community and the engineer in the rubber industry. While a multitude of filler and matrix-related effects need to be taken into account, the most fascinating, but despite its crucial significance for the performance of rubber products, hardly understood phenomenon certainly is the strain-induced crystallization. It leads to the outstanding ultimate and tear properties of natural rubber, making it the material of choice for high-demanding applications, even more than 70 years after the invention of synthetic rubbers. Around 5 million tons of natural rubber annually go into the tire industry, giving natural rubber the largest market share in the elastomer market in terms of total revenue.

Owing to the industrial importance, but also driven by academic motives, natural rubber has been subject to research for a long time. Since the first paper on strain-induced crystallization appeared almost a 100 years ago, numerous studies tried to elucidate the roles of strain, temperature, and filler on the crystallization process. An often overlooked effect is the time, which becomes critical in the context of rubber applications under dynamic load. The majority of natural rubber is consumed for the production of tires that are subjected to highly dynamical loading conditions, with characteristic timescales in the sub-second regime. In order to establish structure–property relationships, structural characterization needs to be carried out on similar timescales. Conventionally, crystallinity is analyzed by wide-angle X-ray diffraction, which usually is a rather time-consuming method.

Fortunately, over the last two decades, synchrotron radiation sources have seen a gain in photon flux by several orders of magnitude. Along with this progress, new detectors have become available owing to the progress in the semiconductor industry. These developments paved the way for the experiments that are presented in this book. Besides this, the work builds on the group's experience with in-situ tensile experiments at synchrotron sources and the invaluable support by IPF Forschungstechnik, enabling novel and purposefully designed experimental setups. Thus, after the commissioning of the PETRA III storage ring at DESY in fall 2010, it was for the first time possible to study the strain-induced crystallization in real-time under realistic dynamic cyclic loading conditions at a frequency of 1 Hz. Strain-jump experiments allowed new insights into the crystallization kinetics on previously inaccessible timescales in the millisecond range. A kinetic law was established, showing that crystallization proceeds on timescales much longer than the typical loading timescales encountered in various rubber products. This teaches the rubber engineer that frequency and time are extremely critical parameters in the mechanical and structural behavior of natural rubber, and that quasistatic experiments can only serve as a starting point for the estimation of the product performance. Indeed, in-situ observations of the crystallinity around a crack tip under dynamic cyclic load established relations between the crystallization kinetics and the tearing behavior. In addition to the practical implications, the experimental findings also call for sophisticated theoretical approaches in order to understand the crystallization behavior from a molecular viewpoint.

Apart from the strain-induced crystallization, other neighboring fields of rubber research were treated in the project *On-line Structure Characterization of Elastomers During Deformation and Fracture* in the framework of the DFG-funded research group *Fracture Mechanics and Statistical Mechanics of Reinforced Elastomeric Blends* and are presented in this book as well. The orientation of filler particles in carbon black and model filler compounds was studied by ultra small-angle X-ray scattering. In-situ tensile experiments in a scanning electron microscope complemented the reciprocal space investigations.

In addition to the above outlined topics, the present book gives a well-balanced treatment of the underlying theory, an extensive literature overview, and a concise description of the experimental methods.

We hope that the book provides an essential step towards a deeper understanding of the structural and mechanical behavior of rubber and promotes fundamental as well as applied research in this field.

Dresden, April 2014

Prof. Dr. Gert Heinrich
Dr. Konrad Schneider

Acknowledgments

The work presented in this thesis was carried out at the Leibniz-Institut für Polymerforschung Dresden e.V. from 2009 to 2013 in the Department of Mechanics and Structure.

At this point I would like to express my gratitude to some persons, without whom the studies would not have been possible.

I would like to thank Prof. Gert Heinrich for the scientific supervision of the work and the straightforward collaboration. Moreover, being the co-organizer and spokesman of the DFG research group FOR 597 *Fracture Mechanics and Statistical Mechanics of Reinforced Elastomer Blends*, he opened the doors for me to meet with numerous colleagues from related fields.

I thank Dr. Konrad Schneider for mentoring the work and his continuous flow of good ideas. He has always had an open door for me. Thanks to his experience in synchrotron experiments and proposal writing, beamtime applications almost always turned out successful. After all, the synchrotron experiments constitute the fundament of this thesis.

I would like to thank Prof. Norbert Stribeck (Universität Hamburg) and Prof. Erwan Verron (Ecole Centrale de Nantes) for reviewing the thesis.

Thanks go to DESY and the Helmholtz Association for granting beamtime at DORIS II and PETRA III. I extend my thanks to the teams at the BW4 and MiNaXS beamlines for a great and fun time with challenging experiments and fascinating results in an inspiring atmosphere: Dr. Jan Perlich,[1] Dr. Stephan Roth, Dr. André Rothkirch, Dr. Adeline Buffet and Dr. Gonzalo Santoro.

Special thanks go to the IPF Forschungstechnik for the design and construction of various parts of the experimental setup (Dr. Michael Wilms, Mathias Ullrich, Vincent Körber, Dirk Zimmerhäckel and team). It is a real advantage in the scientific competition to be able to perform experiments on excellent in-house designed equipment, without which numerous experiments would have been impossible. Dr. Wolfgang Jenschke even joined us for some beamtimes to take care of the miniature tensile machine directly at the beamline.

I would like to thank Maria auf der Landwehr for the SEM analysis and for making the SEM available for in-situ tensile tests. I appreciate the assistance of the

[1] "If I can help you, it is a pleasure for me to go running to the beamline at 4 in the morning."

colleagues from the mechanics lab and the elastomers group (in particular Dr. Thomas Horst, Dr. Amit Das, Holger Scheibner, Rene Jurk). I thank Dr. Robert Socher for the conductivity measurements.

I pleasantly recall the various extended relaxing lunch breaks with Andi, Jens, Maria, Michael, Niklas, Robert, Sven, Tobias, Ulli, preferably in the sun on the patio. These gave me the energy for some overlong afternooons.

I would like to thank the undergrad research assistants Borris Köpper, Robert Tonndorf, and Fabian Richter and the RISE exchange student Laura Poland for their interest in my work and for their support.

I highly appreciate the administrative support by Gudrun Schwarz, smoothing out any bureaucratic issues and organizing various business trips to DESY, The Netherlands, Belgium, Luxemburg, France, Poland, Ireland, and India.

Being part of the research group FOR 597, I acknowledge the financial support of the German Research Foundation (DFG). I thank the research group colleagues for the frequent and lively exchange of ideas and results in an informal atmosphere, leading to a steep learning curve in the fields of mechanics and elastomers.

Finally, I would like to express my thanks to all colleagues in the Department of Mechanics and Structure and the whole IPF for a great time. Furthermore, I would like to thank Prof. Oskar Nuyken (TU München), who for the first time put me in contact with polymers in his macromolecular chemistry class in 2006 and ignited my enthusiasm for polymer research, which has never left me since.

Ultimately, huge thanks go to my parents,[2] who always supported me to pursue my ambitions.

[2] Whose well equipped workshop let me build a dynamic cyclic in-situ tensile machine within a day.

Contents

Symbols and Abbreviations

α	Azimuthal angle
α	Stretch ratio
ΔF	Helmholtz free energy
$\in$	Strain
λ	Wavelength
Φ	Degree of crystallinity
ϕ	Filler volume content
σ	Stress
Θ	Half of the angle between the incoming and scattered beam
A	Strain amplification factor
b	Length of a statistical chain segment
c	Crack length
D_m	Mass fractal dimension
D_s	Surface fractal dimension
E	Tensile modulus
f	Force
h	Height
I	Intensity
k	Boltzmann constant ($= 1.3806 \times 10^{-23}$ J/K)
l	Length
l_0	Undeformed length
M	Mass
N	Number of chains per unit volume
N	Number of cycles in tear fatigue
n	Number of statistical chain segments per chain
q	Magnitude of scattering vector $\left(= 4\pi \frac{\sin(\Theta)}{\lambda}\right)$
R	Aggregate size
R	End-to-end distance of a chain
r_0	Primary particle size
S	Surface area
s	Magnitude of scattering vector $\left(= 2 \frac{\sin(\Theta)}{\lambda}\right)$
T	Tearing energy

T	Temperature
t	Thickness
t_{90}	Vulcanization time obtained by a vulcameter test
w	Strain energy density
w_{el}	Elastic strain energy density
DEA	Dielectric analysis
DMA	Dynamic mechanical analysis
DSC	Differential scanning calorimetry
E-SBR	Emulsion SBR
IR	Infrared
IR	Synthetic isoprene rubber
NMR	Nuclear magnetic resonance spectroscopy
NR	Natural rubber
phr	Parts per hundred rubber (by weight)
S-SBR	Solution SBR
SANS	Small angle neutron scattering
SAXS	Small-angle X-ray scattering
SBR	Styrene-butadiene rubber
SENT	Single edge notched tensile
SIC	Strain-induced crystallization
TEM	Transmission electron microscopy
USAXS	Ultra small-angle X-ray scattering
WAXD	Wide-angle X-ray diffraction

Chapter 1
Introduction

1.1 Materials

Rubbers are chemically crosslinked linear polymers. Their most notable characteristic is their mechanical behavior, i.e. they can withstand large strains and can be loaded multiple times while recovering their initial shape—apart from a small permanent set—upon unloading. Due to the permanent nature of the chemical bonds between the polymer chains, rubbers distinguish themselves from thermoplastic polymers in that they cannot be melted, which implies different processing routes.[1] Compared to thermoset resins, rubbers feature a lower crosslinking density and a higher molecular weight, providing for the lower modulus and higher strain at break.

Historically, the first type of rubber known to mankind was natural rubber. More than 3,000 years ago, native tribes in Central America used rubber balls obtained from the latex of the hevea plant for their now-famous ball games [1]. The next big step in the history of rubber is due to Charles Goodyear, who discovered the process of sulfur vulcanization in 1839. This made rubber products suitable for a wide variety of applications, which were previously excluded from the use of unvulcanized rubber due to the poor mechanical performance and stability. The first synthetic rubbers were developed in the 1930s and became available on an industrial scale in the 1940s, when the supply of natural rubber was limited due to the Second World War. Nowadays, numerous types of rubbers are available on the market, each serving certain purposes. Rubbers are widely used in applications such as tires, dampers, tubes, seals, belts, cable coatings, footwear, textiles, and others.

[1] Sometimes thermoplastic elastomers are also classified as rubbers. These materials are melt-processable and are physically crosslinked rather than chemically.

K. Brüning, *In-situ Structure Characterization of Elastomers during Deformation and Fracture*, Springer Theses, DOI: 10.1007/978-3-319-06907-4_1,

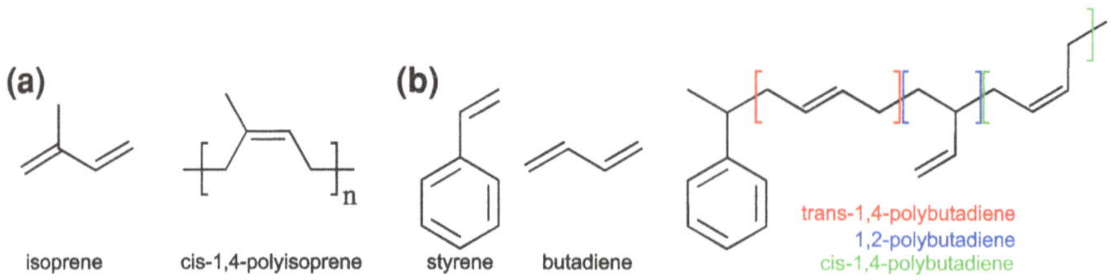

Fig. 1.1 Chemical structures of **a** isoprene and cis-1,4-polyisoprene; **b** styrene, butadiene and its types of appearance in SBR

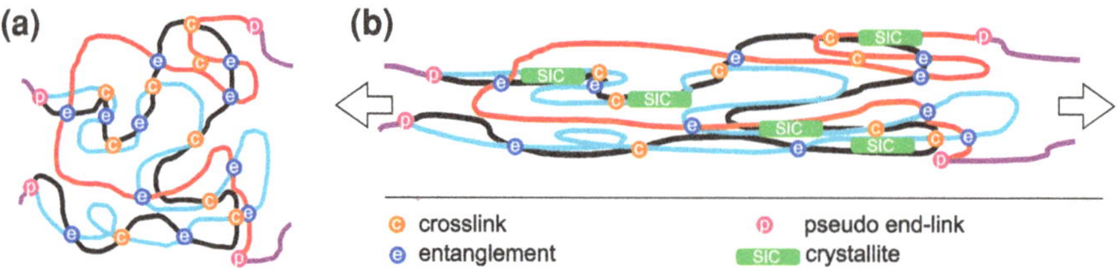

Fig. 1.2 Sketch of network structure of natural rubber: **a** undeformed, **b** stretched beyond SIC onset (Adapted with permission from Ref. [4], copyright 2013 American Chemical Society)

1.1.1 Rubber Types

Natural rubber is obtained from the latex of the rubber tree *Hevea brasiliensis*.[2] After the latex has been harvested from the bark of the tree, it is purified and pressed. The drying process takes place either in air or in smoke. The smoking process conserves the rubber and serves as a protection against fungi. While the smoked version is more relevant on an industrial scale, the air dried version is preferred in science due to its purity (e.g. pale crepe grade, Fig. A.6).

Chemically, natural rubber consists mainly of cis-1,4-polyisoprene (Fig. 1.1a). The perfect stereoregularity (100 % cis) renders NR crystallizable, either by cooling in the quiescent state or by stretching. Besides the polyisoprene, NR contains around 6 % impurities, among which the biggest percentage share the lipids, proteins and other low molecular weight carbohydrates [2]. The polyisoprene chains in NR are linear, but end-functionalized, producing a three-dimensional network with proteins and lipids acting as crosslinkers (Fig. 1.2) [3, 4]. This network gives the comparatively high strength to the unvulcanized rubber, the so called green strength.

In contrast to NR, synthetic isoprene rubber (IR) is not perfectly stereoregular, having a trans content of at least 1.5 % [5, 6]. The irregularity impedes crystallization, shifting the onset strain of SIC to larger strains as compared to NR [7–9]. This is reflected in inferior tensile properties and a reduced tear resistance [10]. Due to the absence of a network in the unvulcanized (green) IR, the green strength is lower by a

[2] Other sources, like dandelion and guayule (*Parthenium argentatum*), have been explored, but are not significant in practice.

factor of 60 [5]. IR is mainly used as a processing aid to facilitate mixing of NR [6]. The yearly consumption of IR is around 20 times less compared to NR [11].

The third type of rubber studied in this work is styrene-butadiene rubber (SBR). Chemically, it is a statistical copolymer of styrene (commonly around 25 %) and butadiene (around 75 %) (Fig. 1.1b). The polymerization is carried out either anionically in solution or radically in emulsion. The first gives solution SBR (S-SBR), the latter emulsion SBR (E-SBR). The type of polymerization determines the vinyl content,[3] which is higher in S-SBR. SBR has almost the same tensile strength as NR, and this at much lower cost. Therefore, it has replaced NR in some applications, notably in passenger car tires. Both S-SBR and E-SBR are used in tires. However, the tear resistance of SBR is inferior to NR, since SBR does not strain-crystallize.

Other rubber types include, amongst others, polybutadiene, polychloroprene, ethylene-propylene rubber and butyl rubber. They find use due to specific properties regarding mechanical hysteresis, gas permeability, chemical stability, oil resistance, heat resistance, ozone aging, flame retardance, glass transition temperature or adhesion [6].

1.1.2 Vulcanization

In the vulcanization process, the chemical crosslinking reaction takes place and the rubber part obtains its final shape. Two types of crosslinking systems are relevant, and both have been used in this work:

- sulfur vulcanization
- peroxide vulcanization.

Industrially, the sulfur vulcanization is more widely used. Sulfur is added in elemental form during the mixing process. Upon curing, it attacks the unsaturated bonds in the backbone or at side groups of the polymer chain and forms sulfide bridges. Depending on the vulcanization system, various types of sulfide bridges with different lengths can be formed. Without the proper use of additives, this crosslinking reaction would take several hours [12].[4] Zinc oxide and stearic acid are usually added as activators, and organic compounds, bearing a $N{=}C{-}S_2$ functionality, serve as accelerators [13]. The precise role of these chemicals in the crosslinking mechanism is still subject to research [14]. The common understanding is that stearic acid aids in the solubilization of the zinc oxide, which then forms a complex with the accelerators and activates the sulfur [15].

Industrial rubber recipes often contain several accelerators or boosters. Retarders are added to extend the scorch time to avoid premature curing during processing.

[3] Vinyl groups result from the 1,2-polymerization of butadiene, in contrast to the more common 1,4-polymerization.

[4] Own tests showed that a standard formulation of natural rubber, including 1 phr sulfur, 1 phr stearic acid and 1.5 phr CBS (accelerator, N-cyclohexyl-2-benzothiazole-sulfenamide), but not containing any zinc oxide, would not vulcanize within 1 h at 160 °C.

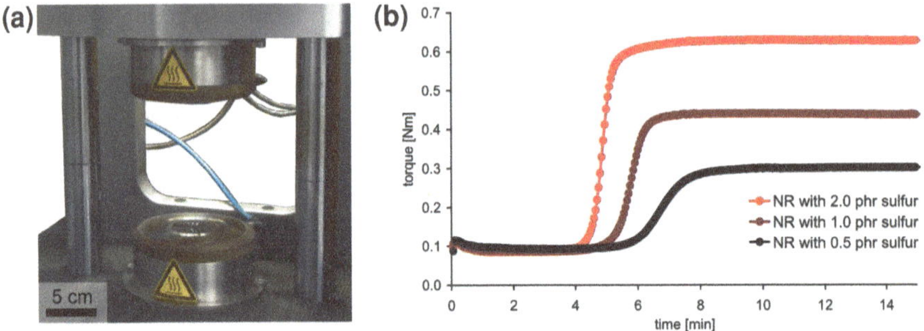

Fig. 1.3 **a** Photograph of the vulcameter dies, between which the sample is placed for analysis. **b** Vulcameter curves (torque over time) at 160 °C from NR compounds with various crosslinking densities (based on #2 (see list of materials in Sect. A.6), keeping the sulfur-CBS ratio constant)

Peroxide vulcanization, on the other hand, is mainly used in a scientific context in model compounds. The chemistry is simpler and no addition of particulate additives like zinc oxide is required. The peroxide decomposes into two radicals at the vulcanization temperature and abstracts a hydrogen from a saturated carbon–carbon bond. The resulting polymer radicals combine to form a direct crosslink without a bridge inbetween [12].

The crosslinking reaction is usually carried out at temperatures between 120 and 180 °C. Following the Arrhenius equation, the curing process is accelerated by increasing the temperature, but care must be taken to avoid thermal degradation. The vulcanization time is measured by a vulcameter, also called rubber process analyzer (Fig. 1.3a). A green rubber sample is placed between two dies, which have been heated to the vulcanization temperature. The dies are firmly pressed together to avoid wall slip and one of the dies starts a sinusoidal rotating motion, subjecting the rubber sample to a shear stress. The torque required to apply a certain shear strain amplitude is taken to be proportional to the crosslinking density, following the simple relation $G = NkT$[5] [16]. After some time, the vulcanization process is complete and the torque reaches a plateau. The time it takes to reach 90 % of the plateau level is commonly referred to as t_{90} and taken as vulcanization time. Depending on the geometry of the mold and the thickness of the rubber part, some additional time might be added to this value.

1.1.3 Fillers

In order to meet the requirements imposed by numerous applications, almost all rubber parts consist of filled rubbers, i.e. they are a composite of the vulcanized

[5] G is the shear modulus; N is the number of chains per unit volume; k is the Boltzmann constant; T is the absolute temperature.

Table 1.1 Typical tire tread compounds for passenger car and truck tires. Additives are not listed. The values are given in phr. Adapted from [17]

	Passenger car tire tread	Truck tire tread
E-SBR	65	
BR	35	
NR		100
Carbon black	70	50
Processing oil	40	10

rubber matrix and a filler. The reason for the addition of a filler can be twofold: either a reduction of the product cost, or an improvement in the (mechanical) properties. The first group is called non-reinforcing fillers, or extenders (e.g. calcium carbonate and other micron-size particulate fillers), the latter reinforcing fillers [12]. The most important reinforcing fillers are carbon black and precipitated silica. They enhance the modulus, increase the tensile strength by a factor of up to 10, and also benefit the abrasion, tear and damping behavior [6].

In rubber technology, the filler content is specified in parts per hundred rubber (phr) by weight, i.e. the rubber matrix constitutes 100 phr by definition. In a blend of different elastomers, the sum of all elastomer matrix components adds up to 100 phr [17]. Typical recipes for passenger car and truck tire treads are listed in Table 1.1. The carbon black loading is between 40 and 90 phr. In passenger car tires, SBR is preferred over NR due to its lower price. In truck tires, however, due to the high loads acting on the tire and due to the requirement for increased mileage, NR is preferred over SBR, e.g. to avoid failure by so-called chipping and chunking.

Over the last two decades, carbon black has increasingly been replaced by silica in passenger car tires, especially winter tires. In truck tires, hybrid filler systems, featuring silica and carbon black, have been introduced.

Processing oil is added to facilitate the processing by lowering the viscosity. It also serves to adjust the hardness and to compensate for high carbon black loadings.

Carbon blacks for reinforcing purposes are produced in the so-called furnace process: Aromatic oil is injected into a fast stream of hot (1200–1900 °C) combustion gases (from the combustion of other carbohydrates), and subsequently quenched with water. Afterwards it is filtered, pelletized and dried. Between the injection and the quenching a partial oxidation takes place due to the controlled presence of oxygen, which is adjusted to substoichiometric concentrations [18]. Within milliseconds, the primary particles are formed from nuclei of solid reaction products, which then fuse together to form aggregates. Besides growth by collision of primary particles, also a gradual deposition of other carbonaceous reaction products on the surface of primary particles and aggregates takes place, acting like a glue between the primary particles and increasing the stability and compactness of the aggregates [19, 20]. Aggregates typically consist of 8–60 primary particles [21–25]. These aggregates cannot be broken up in subsequent processing steps (e.g. rubber compounding). Adjustments in the reaction temperature, reaction time and flow turbulence determine the particle size and aggregate structure [18].

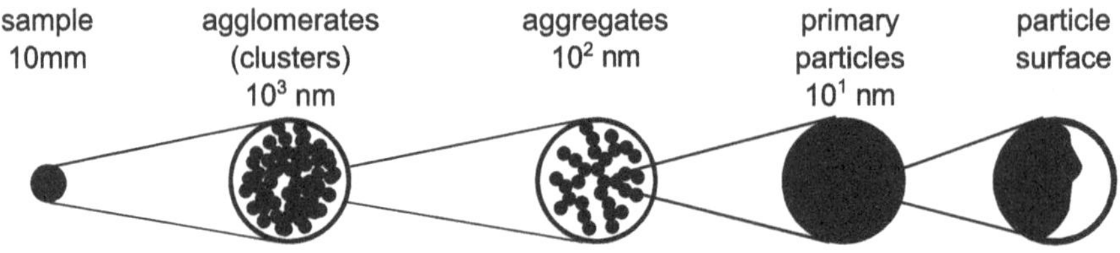

Fig. 1.4 Schematic representation of the hierarchical structure in a carbon black filled rubber (Adapted from Schneider [27])

Carbon black is available in numerous grades for specific purposes, classified by surface area and structure. The standard classification follows the N*abc* nomenclature according to ASTM D-24 [17]. *N* stands for *N*ormal curing.[6] The first digit, *a*, refers to the diameter of the primary particles, which is, as a rule of thumb, 10 *a*, measured in nm, i.e. an N300 grade has a primary particle diameter around 30 nm. The other two digits, *b* and *c*, are assigned based on the structure and surface area. The packing density and the degree of branching within an aggregate determine its structure, which is characterized by dibutylphthalate absorption,[7] while the surface area is analyzed by nitrogen, iodine or CTAB[8] adsorption [17, 18]. Depending on the size of the probing substance, it penetrates into pores and other smaller entities, thus lending the method a sensitivity in a certain size range [26].

The architecture of carbon black is schematically shown in Fig. 1.4. Typically during the processing of the rubber composite, aggregates cluster together to form agglomerates, which are more loosely bound. On an even larger length scale, provided the filler content is sufficiently high, a network superstructure can be established.

The description of the complex carbon black structure in mathematical terms is rather daunting. The most widely used concept is the fractal theory. The foundations for this theory were laid by Benoit Mandelbrot in 1975, describing self-similar geometries. A geometry is self-similar, when each part of it is constituted by a geometric reduction of the whole [28, 29]. This implies scaling laws for geometric properties like the volume (mass) and surface area. Following fractal theory, the area of a rough surface depends on the length r of the yardstick that one uses to measure the area S:

$$S(r) = r^{2-D_\mathrm{s}}. \tag{1.1}$$

A smooth surface has a fractal dimension D_s of 2, whereas for an infinitely rough surface $D_\mathrm{s} = 3$.

Applying the fractal concept to three dimensions, the relation for mass fractals reads:

$$M(r) = r^{D_\mathrm{m}}. \tag{1.2}$$

[6] Nowadays, all industrially relevant standard carbon blacks for mechanical reinforcement are of *N* type, but historically other grades were common.

[7] OAN: oil absorption number.

[8] Cetyltrimethyl ammonium bromide.

Accordingly, a space-filling solid body has a mass fractal dimension D_m of 3, and with increasing pore volume the mass fractal dimension decreases.

Theoretically, the self-similarity and thus the scaling laws 1.1 and 1.2 extend over all size ranges. In practice, there are upper and lower limits. For instance, in the case of fractal aggregates, the scaling holds only for $r_0 < r < R$, where r_0 is the size of a primary particle and R is the size of the aggregate. The fractal dimension of aggregates can actually be predicted by theories. Depending on the growth mechanism, whether it is diffusion-limited or reaction-limited, whether growth takes place by the addition of primary particles to the aggregates or by the clustering of existing aggregates, different fractal dimensions D_m are obtained. They range from 1.75 to 2.2 [29]. As pointed out above, the reality is not as simple as the theory. Due to the inhomogeneity of the reaction conditions along the gas jet in a furnace reactor, several growth mechanisms might occur. The addition of carbonaceous material even smaller than the primary particles adds to the complexity of the growth mechanism. This is further exacerbated by the polydispersity. While the polydispersity of the primary particles is quite low, the aggregate sizes cover a rather broad range, obscuring a clear cutoff size of the scaling region. Not only are the aggregates polydisperse, they have also been observed to be constituted of a rather small number of primary particles. This restricts the scaling law to a rather small region, which rarely covers even a single decade.

In fact, even though fractal theory is the most widely used approach for the description of carbon black geometries, it is not undisputed in the research community. The popularity of the fractal concept in the field of carbon black rather seems to stem from the simplicity of the fractal concept and from the lack of alternative theories rather than from its ability to precisely account for the experimental observations. This conflict is nicely expressed by Huber and Vilgis, who in one of their papers, dealing with carbon black aggregates, use the fractal theory "[...] only to describe the structure, rather than to suggest that the structure *is* fractal" [30].

The beauty of the fractal concept is that it is easily applicable to scattering data, which is one of the favorite methods to characterize carbon blacks. Due to the sizes involved, scattering takes place in the SAXS and USAXS regimes. Bale and Schmidt derived a power law relation between the scattering intensity I and the scattering vector q (Sect. 1.2.2) [31, 32]. For mass fractals, the fractal dimension D_m can be obtained directly from the negative slope of a double logarithmic scattering plot, which is obtained from a slice of a two-dimensional isotropic scattering pattern:

$$I(q) = q^{-D_\mathrm{m}}. \tag{1.3}$$

Equation 1.3 holds in the region $1/R \ll q \ll 1/r_0$. For surface fractals, the following relation holds in the region $q \gg 1/r_0$:

$$I(q) = q^{6-D_\mathrm{s}}. \tag{1.4}$$

In the field of scattering, the polydispersity problem is aggravated by the fact that the small angle scattering intensity does not represent the number average of scatterers, but rather the seventh moment of the number density distribution, i.e. larger scatterers contribute very significantly [33]. Martin showed that if the polydispersity follows a power law distribution, which is a well-founded assumption, the dependence of the scattering intensity of the surface fractal dimension vanishes [34]. This was later confirmed by Beaucage and Hjelm [35, 36]. Both argue that the surface fractal regime is only apparent, but in reality is caused by polydispersity. Ruland backs up this claim, listing numerous published electron microscopy studies, proving that the surface of carbon black is smooth [37], while Beaucage points out that different electron microscopic methods yield different surface roughnesses [26]. Heinrich and Klüppel developed a model to describe the structure of carbon black based on its formation [38].

Beaucage developed a unified model, incorporating several scattering regimes into one equation. The contributions from the Guinier regimes and fractal regimes (Sect. 1.2.3) are weighted by physically motivated scalars and are superimposed using a set of empirical error functions, which secure that each scattering term only contributes to the corresponding regime [39]. While the unified model removes some of the ambiguity in the rather arbitrary individual fitting of the power law regions and crossovers, it provides only limited novel physical insights due to the partially empirical nature of the equation.

It should be mentioned, however, that independent methods also suggest a fractal structure. Klüppel demonstrated that the number of adsorbed gas molecules on a carbon black surface relates to the adsorption cross section of the molecule following a power law scaling [40]. Assuming a monolayer adsorption mechanism, this suggests a fractally rough surface in the probed size range. Schröder showed that the adsorption not only depends on the surface topography, but also on the distribution of adsorption sites of various energy levels [41].

A dependence between aggregate mass fractal dimension and filler loading is to be expected due to the presence of agglomerates [42], introducing crossterms in the scattering relation. This has indeed been observed [43]. On the contrary, a change in surface fractal dimension with filler loading, as observed by Fröhlich et al. [44], is beyond what is expected by theory. Also the broad ranges of values and the lack of agreement between different publications confirm the impression that current theory is insufficient to precisely describe the reality. Some of the experimentally obtained fractal dimensions even lie out of the physically possible range, e.g. mass fractal dimension less than unity are not physically sound [43].

For instance, the mass fractal dimensions D_{m} of N330 obtained through SAXS by different groups vary over a broad range between less than unity and 2.0 [43, 44]. TEM (transmission electron microscopy) measurements of the same carbon black grade yield $D_{\mathrm{m}} = 2.4$ [45]. Theoretically, assuming a cluster-cluster aggregation process for the formation of aggregates in the furnace reactor, carbon black aggregates should have a mass fractal dimension D_{m} of 1.78 [40, 46].

The situation is further complicated by the fact that the classical fractal concept solely describes isotropic structures. Despite the presence of anisometric carbon black aggregates (e.g. ellipsoidal or linear [45, 47]), the isotropy of the scattering patterns seems to suggest the applicability of fractal theory. However, the isotropy only stems from the non-preferential orientational distribution of anisometric objects. Therefore, the application of isometric concepts on length scales of the aggregates can only serve as a rather coarse and simplified characterization of the real structure. This limitation becomes obvious when the black-filled rubber is deformed. Anisometric objects orient along the principal strain axis, and thus the scattering patterns become anisotropic. The classical fractal theory is not applicable under these conditions. Instead of using self-similar concepts, self-affinity can now be postulated, i.e. *each part is a linear geometric reduction of the whole but with ratios depending on direction* [29]. In other words, a self-affine fractal becomes self-similar, when it is scaled with a constant along at least one of the dimensions. This defines a self-affine mass fractal as follows:

$$M(ax_1, ax_2, a^H x_3) = a^{D_m} M(x_1, x_2, x_3), \tag{1.5}$$

where M is the mass, x_i are the spatial dimensions, a is an arbitrary scalar, and H is the Hurst exponent [27]. Other definitions exist. Self-similarity is a special case of self-affinity ($H = 1$). Several authors pointed out that, due to the introduction of the Hurst exponent H, the fractal dimension of a self-affine fractal is not uniquely defined. In fact, very different self-affine structures can have the same fractal dimension [48, 49].

1.1.4 Reinforcement Mechanisms

In general terms, *reinforcement* is the improvement of material properties. More specifically, rubber materials can be reinforced in order to exhibit an increased stiffness, modulus, rupture energy, tear strength, tensile strength, fatigue resistance and/or abrasion resistance [50]. The reinforcing mechanism in filled rubbers is still not fully understood [19, 51, 52], but numerous literature is available pointing at the contributions of several factors.

Examples for the enhanced tensile properties by filler reinforcement are shown in Fig. 1.5. Payne showed, that the small-strain storage modulus of butyl rubbers can be enhanced by as much as 200 times by the addition of carbon black [53, 54].

The filler effect is less pronounced in strain-crystallizing rubbers, such as natural rubber, due to the inherent self-reinforcement (Fig. 1.5b).

In the following, the three main factors of filler reinforcement are briefly summarized.

1.1.4.1 Hydrodynamic Strain Amplification

The basic concept of strain amplification is that due to the rigidity of the filler phase (in contrast to the soft rubber matrix), the rubber matrix undergoes larger deformations than the external deformation would suggest. Gehman expressed this nicely: "[In

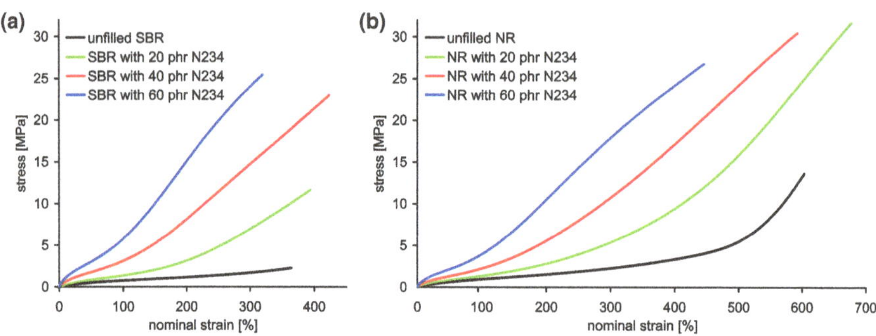

Fig. 1.5 Stress-strain curves for **a** SBR [#40, #41, #42, #43] and **b** NR [#2, #3, #4, #5] compounds with various filler loadings. The reinforcing effect is clearly visible (For sample nomenclature, see Sect. A.6)

filled rubbers,] rubber works at a higher point on its stress-strain curve" [55]. In the context of a viscous particle suspension, the physics behind this effect have been worked out by Einstein more than 100 years ago. He proposed an equation relating the viscosity of the suspension η to the amount of filler volume content ϕ [56]. For spherical particles, the equation reads

$$\eta = \eta_0 \left(1 + 2.5\phi\right), \tag{1.6}$$

where η_0 is the viscosity of the unfilled fluid. Equation 1.6 is only valid for non-interacting spheres, i.e. the filler content must be sufficiently low. Einstein's relation was applied to the modulus E of filled rubbers by Guth and Gold and extended for a correctional quadratic term, taking care of the filler interaction in concentrated systems:

$$E = E_0 \left(1 + 2.5\phi + 14.1\phi^2\right). \tag{1.7}$$

The quadratic term is empirical and other factors have frequently been suggested [52, 56]. The equations for hydrodynamic strain amplification can only serve as a rough approximation of the reinforcement due to neglecting effects like strain rate and strain history [57].

Medalia quantitatively introduced the concept of an effective filler volume [55, 58]. The effective filler volume includes not only the filler itself, but also the passive rubber fraction, i.e. the part of the rubber that is occluded in the aggregate pores and is trapped in the agglomerate structure and thus does not undergo the same deformation as the rubber chains in the bulk phase of the composite. The effective filler volume depends on the structure of the carbon black and its dispersion [20, 26, 40, 59]. Typical values for the ratio of the effective filler volume to the nominal filler volume range from 1.3 to 2.0.

As mentioned above, the strain amplification was originally applied to the modulus of the material, i.e. to the small strain region, which can be approximated by linear elasticity. It is also applicable to the strain in general. According to Bueche and Smallwood, the strain is amplified by a factor of $\frac{1}{1-\phi^n}$, where the exponent n depends on the geometry of the filler particles [56, 60, 61].

Even though the concept of strain amplification is widely accepted, the direct proof by experimental measurement of the matrix strain is difficult. Approaches include mechanical tests [62, 63], NMR (nuclear magnetic resonance spectroscopy) [60, 64], SANS (small angle neutron scattering) [59, 65–67] and WAXD [68, 69]. The underlying assumption is that a unique relation between the measured quantity ψ (e.g. stress, crystallinity, orientation) and strain ϵ exists, which applies to the unfilled material as well as to the matrix of the filled material [63, 68]:

$$A_{\psi,\text{filled}}(\epsilon) = \left.\frac{\epsilon_{\text{unfilled}}}{\epsilon_{\text{filled}}}\right|_{\psi} = \left.\frac{\epsilon_{\text{matrix}}}{\epsilon_{\text{filled}}}\right|_{\psi}, \tag{1.8}$$

where $A_{\psi,\text{filled}}$ is the strain amplification factor of the filled material for the quantity ψ. For instance, let the strain in a filled rubber be half of the strain in an unfilled reference at the same stress, then the mechanical strain amplification factor is two. The concept of the mechanical strain amplification has proven to be useful as long as the reinforcing capability of the filler is high, such as in carbon black filled systems [63], and the strain is larger than the Payne regime, i.e., above 20 %, but below the SIC onset [64]. Given that hydrodynamic effects are not the only source of reinforcement, and also numerous other mechanisms like filler–filler networks and adsorption of polymer at the filler surface (immobilized layer of bound rubber, filler particles as additional network crosslinks) have to be considered, quantitative deviations from a constant strain amplification are to be expected and have indeed been observed [60, 64].

In general, the strain amplification factors obtained by different methods do not agree, because the methods are sensitive to different characteristics. Furthermore, it is often overlooked that the concept of strain amplification is not applicable at high carbon black loadings, since above the overlap concentration of the agglomerates, other reinforcing mechanisms set in (Sect. 1.1.4.2) [20]. Some SANS studies did not find any overstrain in the matrix [66, 67].

1.1.4.2 Filler–Filler Interaction

At filler loadings below the overlap concentration, no filler network exists and the load transfer takes place through the rubber matrix, implying the strain amplification as pointed out in the previous section. On the contrary, above the overlap concentration, a filler network is established, and the stresses are transferred directly via the agglomerates. Since the modulus of this network is much larger than that of the rubber, the filler network is the dominating element in this regime [20]. The existence of the filler network is the reason why aggregated structures have better reinforcing properties than completely dispersed fillers [70].

1.1.4.3 Filler Surface and Interphase

The processes at the interface between the rubber chains and the carbon black filler surface is still subject to current research, even though the notion seems to prevail that the interaction is mainly physical in nature, i.e. physisorption and Van-der-Waals interaction dominate over chemical bonding [51, 71]. Applying the macroscopic concept of surface tension and internal pressure to the particle interface, adhesive failure at the filler surface is hindered by these factors alone as long as the particles are sufficiently small, i.e. below 50 nm [51]. Indeed, it is observed that the reinforcing effect increases with decreasing particle size, and that above a certain particle size the reinforcing effect vanishes. The minor role of chemical interaction was confirmed by placing different amounts of reactive groups along the polymer chains. Despite the presence of more reactive groups, the polymer-filler interaction was not enhanced [51]. The idea of physical interaction implies that chains are able to slide along the filler surface, causing hysteresis [51].

Due to the physical adsorption of the polymer chains on the carbon black surface, their mobility is restricted, resulting in an immobile layer, also called glassy layer. The presence of this layer is suggested by DEA (dielectric analysis) [40, 72], DMA (dynamic mechanical analysis) [50], NMR [73], chemical extraction [74] and 3D TEM [75, 76], even though direct evidence is still lacking [67, 71]. The restriction of the polymer mobility on the filler surface is seen in analogy to thin films deposited on substrates, where an increase in the glass transition temperature T_g of the order of 100 K is observed [54]. A more detailed model describes the immobile layer in terms of a gradual transition from a highly immobilized state directly at the surface to bulk-like mobility further away from the surface [77]. The thickness of this transition region is said to range between 1 and 10 nm.

The role of the interphase becomes increasingly important as the particle size is reduced and the specific surface is increased. It further contributes to the superior reinforcing ability of smaller particles. While semi-reinforcing carbon blacks have a specific surface of less than 45 m^2/g, reinforcing fillers have around 65–140 m^2/g [17].

1.1.4.4 Self-Reinforcement by Strain-Induced Crystallization

A matrix-inherent reinforcement mechanism is the strain-induced crystallization [78]. Strain-crystallizing elastomers partly crystallize if the exerted strain exceeds a certain threshold. This process is governed by the reduction of enthalpy, which depends on the molecular architecture of the polymer chain. A critical prerequisite for the ability to crystallize is the absence of large sidegroups. A high stereoregularity (cis/trans) is required if the backbone contains double bonds (cf. tacticity effects in vinyl thermoplastics). Natural rubber is the most prominent elastomer that exhibits SIC. Other

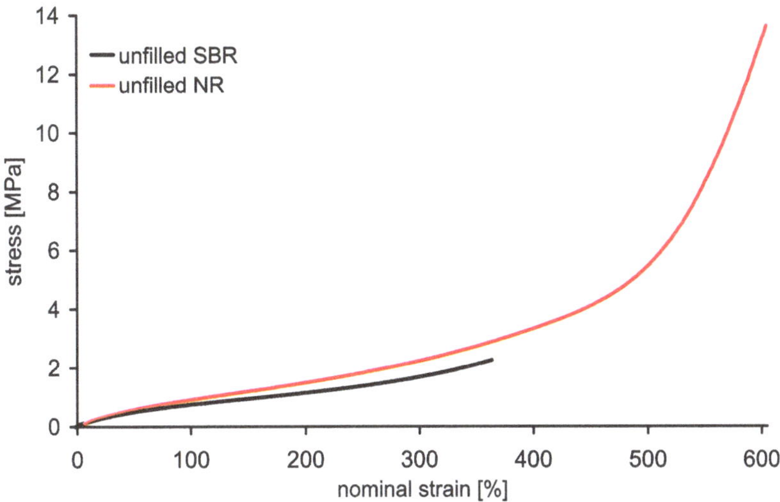

Fig. 1.6 Stress–strain curves for NR and SBR. The stress upturn of NR can be assigned to the reinforcing effect of strain-induced crystallization. The detailed recipes are listed in Sect. A.6 [#2, #40]

strain-crystallizing elastomers include isoprene rubber (IR) [5, 10],[9] chloroprene rubber (CR) [78, 80–83], butyl rubber [84] and butadiene rubber [84].

Crystallites can be thought of as additional crosslinking points or as rigid reinforcing particles, strengthening the otherwise amorphous rubber [85, 86]. Thanks to SIC, the strength at break and the crack propagation behavior of NR are superior to those of non-crystallizing synthetic rubbers (Fig. 1.6).[10] The beauty of the self-reinforcement is that the crystallization process occurs only in those regions which are subjected to large strains, while the rest of the product maintains its elastic properties. Upon removal of the load, the crystallization is reversible. The drawback is the complex relationship between SIC and temperature, time and multiaxial strain fields. At elevated temperatures and on very short time scales SIC is reduced. This makes an implementation of SIC into constitutive models difficult [87, 88].

1.1.5 Mechanical Behavior

Since rubber is mainly used as a structural material, its mechanical properties are of utmost importance. While the basic elastic behavior can be traced back to simple

---

[9] Only IR grades with high cis-content can undergo SIC. Furthermore, IR crystallizes only in the crosslinked state [79].

[10] Of course, the comparison should not be taken too quantitatively. NR and SBR are chemically different, and moreover NR has an inherent network resulting from its natural impurities and endgroup functionalization. However, from a practical point of view, reinforced SBR can often replace reinforced NR in less demanding applications.

entropy elasticity, the problem becomes somewhat more involved when dealing with filled rubbers or strain-crystallizing rubbers.

1.1.5.1 Origin of Rubber Elasticity

As opposed to metals, the elasticity of rubbers is driven by entropy and not by enthalpy. In a stretched rubber, the chain segments can assume fewer conformations, reducing the entropy, and thus increasing the free energy. In other words, the larger the distance between the chain ends, the fewer paths are available to connect the ends by a given number of segments. The increase in free energy is balanced by the work caused by the external force acting on the rubber. For a single freely jointed chain, following Gaussian chain statistics, the relation between the magnitude of the end-to-end vector R and the force f reads as follows:

$$f = \frac{3kT}{nb^2} R, \tag{1.9}$$

where n is the number of statistical chain segments and b is the chain segment length [89].

Extending the concept from a single chain to N chains per unit volume in three dimensions, the change in the free energy ΔF upon uniaxial elongation is, under the assumption of incompressibility [90]:

$$\Delta F = \frac{NRT}{2} \left(\alpha^2 + \frac{2}{\alpha} - 3 \right). \tag{1.10}$$

Here, α is the stretch ratio.[11] At constant temperature, the free energy ΔF equals the work w done on the system. The work is stored as elastic strain energy, following the definition of a hyperelastic material model. Equation 1.10 represents one of the simplest constitutive equations for the mechanics of elastomers. With $NRT = C_1$, C_1 being an empirical constant, it can be identified as the so-called Neo-Hookean law. It is a special case of the more general Mooney-Rivlin constitutive law. Nowadays, numerous constitutive models are available and are an essential basis for any finite element code [91].

The derivative of the strain energy density w with respect to the displacement gives the stress σ:

$$\sigma = NRT \left(\alpha - \frac{1}{\alpha^2} \right). \tag{1.11}$$

[11] The stretch ratio α is defined as $\alpha = \frac{l}{l_0}$, where l_0 is the undeformed length and l is the deformed length. The relation between the stretch ratio α and the strain ϵ is $\alpha = 1 + \epsilon$.

1.1.5.2 Mullins Effect

The Mullins effect describes the history dependence of the large strain mechanical behavior of filled and unfilled elastomers [92]. This stress softening effect is more pronounced in filled rubbers. When stretching a rubber for the first time above the previous maximum strain, filler–filler bonds break and rubber chains slide along the filler surface [51]. In the consecutive stretching cycles, the stress at a given strain level below the previous maximum strain will be reduced (Figs. 1.7a and 4.5). In addition to the filler, viscous matrix effects also play a role. The Mullins effect is most pronounced during the first three loading cycles and vanishes after ten cycles. After very long waiting time, preferentially at elevated temperature, the Mullins effect is reversible [51, 93]. The Mullins effect should be taken into account when dealing with the reproducibility of mechanical experiments. In order to exclude undesired effects in the study of bulk materials, multiple cyclic loading to the maximum investigated strain prior to the experiment is advisable (so-called *demullinization*). In contrast to this, in the context of static crack growth and tear fatigue, the material in front of the crack tip usually was not subjected to excessively large strains before being reached by the crack front.

1.1.5.3 Payne Effect

A peculiar effect in filled rubbers in the low strain regime is the Payne effect. It nicely connects the filler structure with the dynamic mechanical behavior. The Payne effect is expressed as a drop in storage modulus G' as a function of strain amplitude in a strain sweep DMA experiment (Fig. 1.7b) [53]. Between 0.1 and 15 % strain, G' drops by roughly a factor of ten. The magnitude of the drop depends on the filler structure and on the temperature. The underlying physical reason is the breakdown of filler–filler interactions [71]. Sotta et al. postulated the breakdown of glassy bridges between agglomerates [54]. The Payne effect is not found in unfilled rubbers, but it occurs in carbon black-filled liquids [40, 94]. The fact that the Payne effect is less pronounced at high temperatures hints at the temperature sensitivity of the filler bridges; thus supporting the idea of overlapping glassy layers serving as bridges to establish a network [26, 50]. In terms of strain amplification, the Payne effect can be interpreted in that upon the destruction of the filler network, previously trapped rubber rejoins the deformed phase and thus the effective filler volume is decreased [26]. After a sufficiently long waiting time, the material can recover from the Payne effect [95]. The Payne effect has been implemented into models [40, 61].

1.1.5.4 Fracture and Tear Fatigue of Elastomers

Besides the small-strain dynamic properties, the tear and rupture properties of rubbers are highly relevant for the material performance in various applications. In some cases, tear fatigue is the lifetime-limiting failure mechanism. The beginnings of

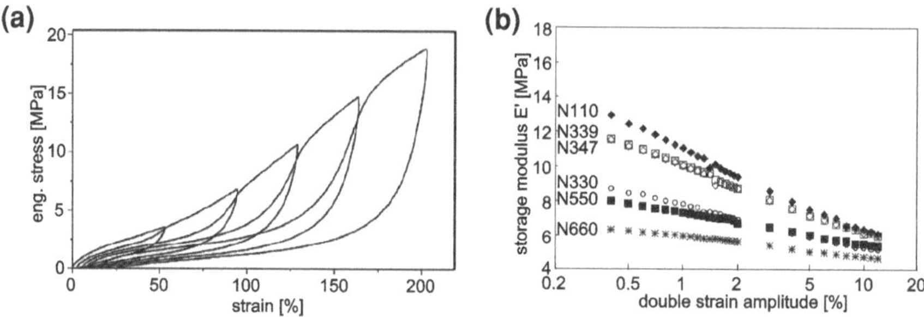

Fig. 1.7 **a** Illustration of the Mullins effect, exemplified on E-SBR with 80 phr N339. (Adapted with permission from Ref. [40], copyright 2003 Springer.) **b** Illustration of the Payne effect in filled natural rubbers with 50 phr of various filler grades. The more reinforcing the filler, the more pronounced the drop in the storage modulus E'. The experiments were performed at 1 Hz with $-10\,\%$ static predeformation at 25 °C (Adapted with permission from Ref. [94], copyright 2002 Springer)

tear fatigue investigations in elastomers can be traced back to the 1940s with the introduction of synthetic rubbers [96, 97]. By definition, the prediction of fatigue life involves the steps of crack nucleation and growth [98, 99]. In the case of rubbers, the common assumption is that cracks propagate from inherently existing defects like filler agglomerates, additives like zinc oxide, impurities occurring naturally in the raw elastomer, and imperfections in mold surfaces [100]. For instance, it was found by SEM that the majority of the cavities contained a zinc oxide granule [101]. Mars even argued that crack nucleation in a narrow sense does not occur due to the presence of flaws [100].

In the early days of rubber fracture research, the Griffith criterion was used for brittle elastic materials, relating the strain energy stored in a notched material to the surface energy γ required to advance the crack. However, this simple concept is not applicable to rubbers, since the stored strain energy cannot be fully converted to surface energy due to dissipation effects [102]. In the 1950s, Rivlin and Thomas extended the energy-based Griffith criterion to elastomers [103, 104]. They introduced the critical energy release rate, which has become the most widely used concept in fracture mechanics of rubbers [104]. They showed that the critical energy release rate, above which the crack propagates, is independent of the sample geometry under certain conditions [97]. In the field of rubber technology, the term **tearing energy** is commonly used for the fracture mechanics expression **energy release rate** [97].

The energetic approach to fracture mechanics can be applied on a global and on a local scale [105]. Applying it globally, the calculation of the tearing energy depends on the geometry of the rubber specimen. Nowadays, the most popular geometry is the so-called **pure shear sample**,[12] because it gives the most accurate results due to inherently better statistics [106]. Other geometries, like the single edge notched

[12] The term *pure shear* might be somewhat misleading. Simply speaking, the pure shear specimen is a tensile specimen with one of its dimensions transversal to the tensile direction being much larger than its extension along the tensile direction, such that transversal strains are restricted to one

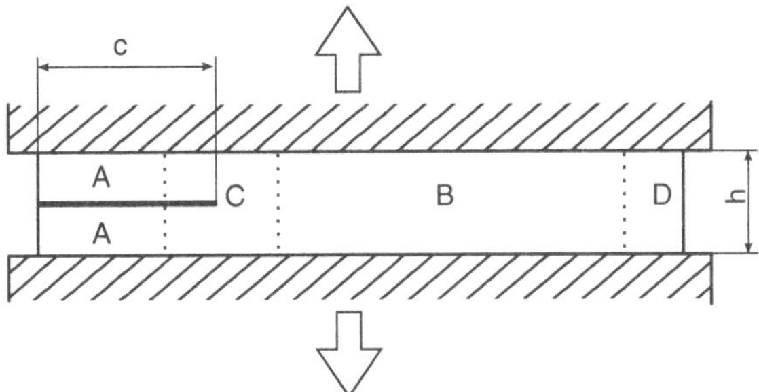

Fig. 1.8 Illustration of a tear fatigue experiment in pure shear geometry. Zone *A* is stress-free, zone *B* is in pure shear state, zones *C* and *D* are in complex states of stress due to crack tip and boundary effects. The *arrows* indicate the tensile direction. Adapted from Ref. [103]

tensile (SENT) sample or the trousers sample, are mainly of historical interest [97, 102, 106]. For a pure shear geometry, the tearing energy can easily be calculated from a global energy balance (Fig. 1.8) [103]. If we let the crack propagate by the distance dc perpendicular to the tensile direction, then the zone C expands in length by dc and zone B shrinks by dc, assuming a self-similar shift of the crack tip. As a first approximation, no change in displacement occurs during the crack growth, i.e. the process is static. Then the energy W transferred from zone B to zone C is

$$W = h\ t\ w_{\mathrm{el}}\ \mathrm{d}c = t\ T\ \mathrm{d}c, \tag{1.12}$$

where h and t are the height and the thickness, respectively, of the specimen; and w_{el} is the elastic strain energy density in the homogeneously deformed zone C. T as the tearing energy follows as

$$T = h\ w_{\mathrm{el}}. \tag{1.13}$$

The elastic strain energy density w_{el} can be obtained from the area under the equilibrium stress-strain curve (which more closely resembles the unloading curve than the loading curve) of an unnotched pure shear sample, possibly with some correction for boundary effects in the D zones. Alternatively, w_{el} is obtained from the strain in connection with a constitutive model. Strictly speaking, Eq. 1.13 is only valid under static conditions, i.e. no work is done on the sample while the crack propagates. To a good approximation, this is fulfilled in quasistatic tear tests, where the sample is stretched continuously and the crack growth is observed. However, these kinds of rupture tests can hardly represent the fatigue loading conditions which are typically encountered in rubber products. Moreover, performing quasistatic tear tests is meaningless for strain-crystallizing rubbers since catastrophic failure occurs once the crystalline zone, acting like a wall, has been overcome [6, 103, 107].

Besides the global approach, a local approach, based on the J-integral, is available. In simple words, this concept, proposed independently by Rice and Cherepanov [108, 109], performs a local energy balance around the crack tip. In analogy to

(Footnote 12 continued)
dimension. The name results from the fact that this deformation field can theoretically be obtained by exerting shear forces on the edges of the sample, along with a rotation of the sample [16].

Green's theorem, the energy components acting on and crossing an arbitrary closed path around the crack tip are summed up to obtain the energy source term within this area, which corresponds to the creation of new crack surfaces and energy dissipation. If dissipation occurs only in the crack tip near field, the J-integral approach becomes path-independent, provided it encircles the crack tip near field. When taking the sample contour as integration path, the J-integral reduces to Eq. 1.13 [110]. If dissipation is not limited to the crack tip near field, the J-integral becomes path-dependent. The analysis of crack propagation in dissipative materials of arbitrary geometry is subject to current research [111].

In practice, cyclic fatigue crack growth experiments are preferred over static tear tests, because they resemble more closely real-life loading conditions. In these so-called tear fatigue tests, the notched rubber specimen is subjected to a cyclic load at a frequency of roughly 1 Hz (pulsed or harmonic) and the crack growth over the number of cycles is observed (Fig. 1.9a).[13] While at small strain amplitudes the growth is dominated by ozone attack (regime 1 in Fig. 1.9b) and at large amplitudes catastrophic failure occurs (regime 3), the region inbetween (regime 2) is characterized by a power law relation between the crack growth per cycle $\frac{\mathrm{d}c}{\mathrm{d}N}$ and the tearing energy T to the power of m:

$$\frac{\mathrm{d}c}{\mathrm{d}N} \propto T^m \tag{1.14}$$

The Paris law (Eq. 1.14) was originally established for metals [97]. m is an empirical parameter and is between 3 and 4 for NR and SBR, respectively [6, 112]. There are several limitations which are not captured by the Paris law. First, the fatigue life not only depends on the load amplitude, but also on the load frequency and load history. Under pulse load, resembling more closely the loading scenario in a tire, the lifetime can be considerably reduced as compared to sine load [113]. The extent of lifetime reduction depends on the compound, such that the material rankings from a sine and a pulse test can be completely different [114]. Also, the introduction of dwell periods or variable amplitude loading histories can drastically change the lifetime by up to factor of ten [115]. Literature does not give an unequivocal picture about the influence of the strain rate and frequency. While Andrews [116] and Gent [6] showed, somewhat counterintuitively, a negligible effect of frequency on crack growth in NR, recent results [113], varying only the strain rate and not the frequency, suggest a decrease in tear resistance with increasing strain rate. However, this could also be related to the inherent variation in dwell time when changing the strain rate at a constant frequency, which was shown to have a large effect [115, 117].

The most influential parameter besides amplitude and frequency is the R-ratio, defined as the ratio of minimum to maximum strain or strain energy in a fatigue cycle [104]. While in amorphous elastomers the R-ratio has a negligible effect [117], in strain-crystallizing NR the crack growth rate can decrease by as much as two orders of magnitude when the R-ratio is increased from 0 to 0.06 [99]. This is nicely

[13] In the framework of the FOR 597 research group, tear fatigue experiment were carried out on a Coesfeld machine at TU Chemnitz.

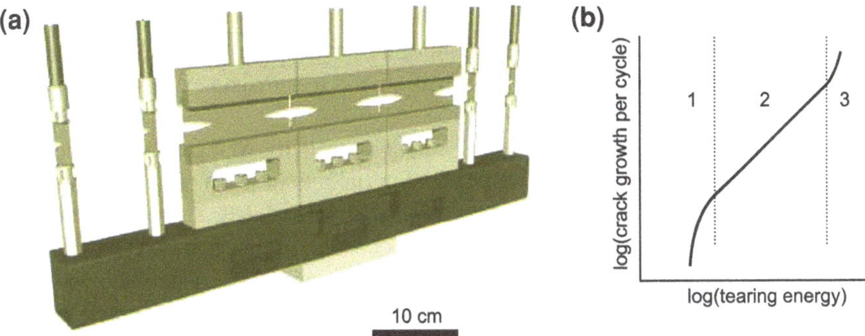

Fig. 1.9 Tear fatigue analysis: **a** Scheme of a tear fatigue analyzer for parallel testing of pure shear and SENT samples (image courtesy of R. Stoček, TU Chemnitz); **b** Illustration of fracture regimes in tear fatigue. *1* minimal crack growth due to ozone attack, *2* power law regime, *3* catastrophic failure (Adapted with permission from Ref. [118], copyright 2012 Springer)

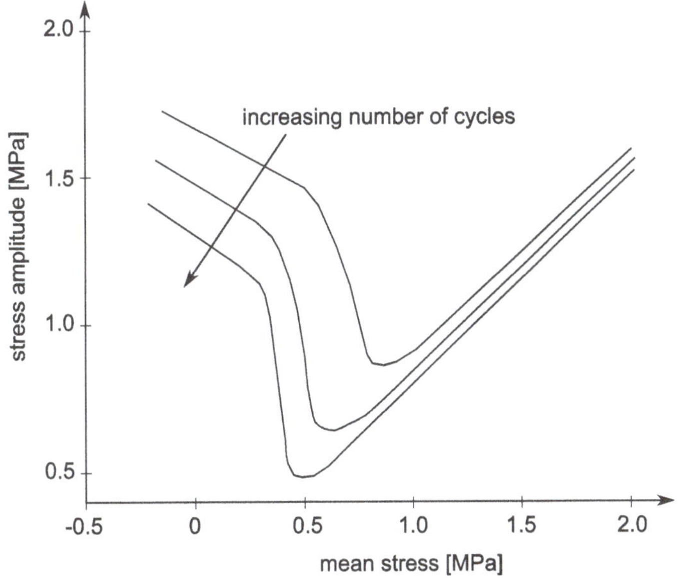

Fig. 1.10 Haigh diagram of filled natural rubber (23 phr carbon black) subjected to harmonic strain at 1 Hz. The criterion for the number of cycles listed in the graph is the presence of a crack larger than 1 mm. The characteristic of natural rubber is the increase of the isolines with increasing positive mean stresses (Adapted from Refs. [121] and [122])

reflected in the Haigh diagram (Fig. 1.10), plotting the stress range of a cycle, $\Delta\sigma$, over the average stress per cycle, $\overline{\sigma}$, which shows a characteristic increase in the isolines for constant lifetime [119], i.e. cycling around a larger strain increases the lifetime despite the higher strain energy density involved, which should theoretically release more energy and favor crack propagation. The reason behind this apparent paradoxon is the strain-induced crystallization, which can develop its reinforcing abilities only when the material is kept above a certain strain for sufficient time. The relation between strain-induced crystallization and tear resistance was first pointed out by Busse in 1934 [120].

The role of SIC in crack growth is also apparent from the topographies of the crack contour and crack surface. In non-crystallizing rubbers or if SIC is suppressed (e.g. due to high strain rates), the crack surface is rather smooth [6]. The more prominent the role of SIC, the more wrinkled the surface becomes. Knotty tearing [123, 124] and characteristic striations [107, 116] result from the reinforcing crystalline zone at the crack tip and the anisotropy of the material in that region, forcing the crack to make a detour. Thus, on top of the large hysteresis associated with strain-crystallizing materials, the real crack surface is much larger than what a simple crack length measurement would suggest, effectively enhancing the apparent tearing energy, e.g. obtained from a pure shear tear fatigue test. Under certain circumstances, even crack rotation and bifurcation are observed in NR [62, 125], effectively increasing the crack tip radius and thus elevating the energy release rate [126]. A more detailed description of the role of SIC and its time dependency follows in Sect. 1.3.2.

The tear fatigue tests outlined above have in common that their loading conditions are limited to uniaxial cases, which contrasts real-life scenarios, frequently exhibiting multiaxiality [127]. Typically, the cracking plane is observed to be normal to maximum principal strain direction [100]. Various multiaxial equivalence fatigue criteria relate the tearing energy obtained from a uniaxial test to the actual strain and stress field under multiaxial conditions [98, 128]. This field is subject to current research [129].

Other factors affecting the fatigue lifetime include the crosslinking type and density [99, 130, 131], and the sample thickness [97, 132, 133].

1.2 X-Ray Scattering

X-ray scattering is a powerful non-destructive analytical technique to probe the nano- and microstructure of materials. Employed in transmission geometry[14] it is bulk-sensitive and, depending on the setup, provides local information or averages over a large volume to yield representative results. X-ray scattering can have high time-resolution in the millisecond range, but can also follow slow processes. X-ray scattering can be done with ease in rather flexible sample environments, e.g. in-situ heating, cooling or mechanical experiments can be implemented.

The data obtained are strictly quantitative, however its interpretation is the main difficulty and drawback of this method. Scattering data represents information in reciprocal space, which is hardly accessible to common human perception, and therefore either a conversion into real space, whether direct or by modeling, has to be done, or, on the assumption of certain geometries, a direct extraction of information from reciprocal space can be performed.

[14] As opposed to grazing incidence geometry.

1.2.1 Nature and Sources of X-Rays

X-rays are electromagnetic radiation, which are located at the high energy end of the electromagnetic spectrum. Their wavelength λ ranges from 0.01 to 10 nm, corresponding to energies between roughly 100 and 0.1 keV. X-rays were discovered in 1895 by Wilhelm Conrad Röntgen.

The most widely used source of X-rays is the X-ray tube, which emits X-rays from a target (typically copper), that is bombarded with electrons. The electrons are emitted from a cathode wire and are being accelerated by an electric voltage. Besides the Bremsstrahlung background, an X-ray tube emits radiation of a few characteristic wavelengths, which result from electrons from distinct outer shells falling down into the vacancies in the inner shell caused by the electron bombardment (fluorescence). For instance, the most frequently used type of radiation in X-ray scattering, Cu Kα radiation, is caused by a transition of electrons from the L-shell to the K-Shell and has a wavelength of 0.154 nm.

The major disadvantage of the X-ray tube is the low efficiency, which is mainly due to the isotropic emission of X-rays from the target. Because a defined beam is required for scattering purposes, only a small portion of the radiation can be used. The low photon flux leads to long exposure times to yield a sufficient signal to noise ratio. This renders time-resolved experiments and high-throughput measurements impossible.

A more powerful alternative is the synchrotron light source (Fig. 1.11). In a synchrotron, electromagnetic radiation is generated from the deflection of charged particles (electrons or positrons) moving at relativistic speeds. The first of these large-scale facilities were built in the 1960s to perform particle physics experiments, and the electromagnetic radiation was only a byproduct. Nowadays, with the third generation of synchrotron sources being available, they are purposefully designed to deliver high brilliance[15] X-ray radiation.

In a synchrotron source, electrons or positrons are accelerated by a linear accelerator and booster and then fed into a storage ring, typically of a few hundred meters in diameter. The particles circulate in the storage ring at relativistic speeds. Being at energies around 5–8 GeV, $1 - \frac{v}{c} \approx 10^{-5}$ (with v being the particle speed and c being the speed of light). In the storage ring, bending magnets force the electrons to move in a circle. The deflection of charged particles causes radiation loss, which is emitted as X-rays. The loss of energy is compensated by acceleration cavities. Besides the bending magnets, other devices (wigglers and undulators) are purposefully inserted into the storage ring to generate more brilliant radiation. Wigglers can be regarded as periodic arrays of n bending magnets, such that the brilliance is $\frac{n}{2} \approx 100$ times higher than that of a bending magnet. Undulators, employed in 3rd generation synchrotrons, are designed similarly as wigglers, but take advantage of amplification effects by interferences between the emitted X-rays from individual magnets, such that their brilliance is again enhanced by two to three orders of magnitude.

[15] The brilliance is the photon flux (i.e. photons per time and per beam cross sectional area) normalized by solid angle (cf. beam divergence or collimation) and 0.1 % bandwidth (cf. monochromacity).

Fig. 1.11 Aerial view of the Petra III storage ring and the Max-von-Laue experimental hall (*blue*) at DESY, Hamburg. It houses the MiNaXS beamline, at which most of the experiments for this work were carried out (Reproduced with permission from Ref. [134], copyright DESY, Reimo Schaaf)

After the X-ray beam has left the storage ring in tangential direction, it passes through some optical devices, which focus the beam (by means of a Göbel mirror or by slits) and confine the wavelength spread of the beam (only waves within a certain wavelength window fulfill the diffraction conditions of the monochromator crystals) and finally arrive at the experimental hutch of the beamline.

The difference in brilliance between a rotating anode lab source and the latest synchrotron sources is more than ten orders of magnitude. This high brilliance is especially useful for USAXS and SAXS. Considering that for WAXD the intensity itself is more relevant, still the exposure times at a synchrotron are reduced by a factor of 10^5 as compared to lab experiments, opening up new methods to study processes on previously inaccessible short time scales.

1.2.2 Scattering and Diffraction

This section deals with X-ray scattering and diffraction as analytical techniques in materials science to probe the nano- and microstructure of polymers. It intends to give a moderately detailed introduction into the theoretical fundamentals of the method, sufficient to understand and interpret experimental and literature data. The section should not be seen as attempt to comprehensively cover the theory of electromagnetic fields or quantum mechanics to deduce scattering theory from first principles.

Along these lines we start with the experimental setup and from there illustrate the physical relations. The theory laid out in the following is based on the classical books by Guinier [136], Feigin and Svergun [137], Glatter and Kratky [138] and a recent publication by Lindner and Zemb [139]. Figure 1.12 shows a sketch of a scattering experiment. Let us assume the incoming beam is a monochromatic planar wave, linearly polarized in the plane of the paper, i.e. the electric field vector points out of the plane of the paper. The electric field of the photon produces an oscillation of electric charges in the sample. As a good approximation, we can neglect the interaction

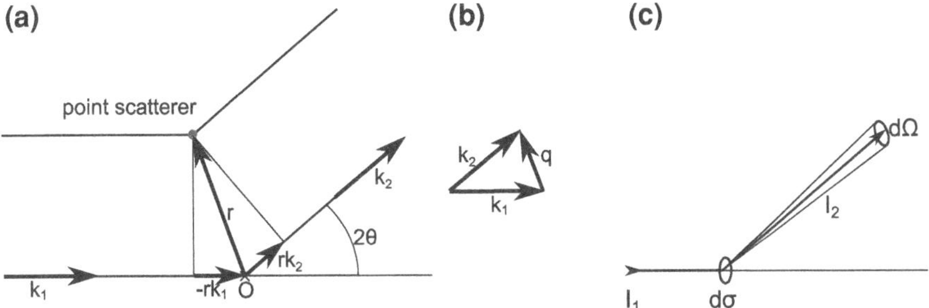

Fig. 1.12 **a** Schematic representation of a scattering experiment. **b** Graphical definition of the scattering vector $\boldsymbol{q}$. **c** Illustration of the scattering cross section $\frac{\mathrm{d}\sigma(\boldsymbol{q})}{\mathrm{d}\Omega}$ (Adapted from Ref. [135])

between X-ray photons and protons due to their large mass.[16] An oscillating electron emits an electromagnetic wave. Assuming elastic scattering ($\lambda_2 = \lambda_1$), the relation between the scattered wave $\boldsymbol{E}_2$ and the incoming wave $\boldsymbol{E}_1$ is

$$\boldsymbol{E}_2 = \boldsymbol{E}_1 \frac{{e_0}^2}{m_0 c^2 r} \exp(-i\boldsymbol{q}\boldsymbol{r}), \tag{1.15}$$

where $r_0 = \frac{{e_0}^2}{m_0 c^2}$ is the electron radius, $\boldsymbol{r}$ is the position of the electron and $\boldsymbol{q}$ is the **scattering vector** $\boldsymbol{q} = \boldsymbol{k}_2 - \boldsymbol{k}_1$ (Fig. 1.12a, b).[17] The magnitude of the scattering vector is, with $|\boldsymbol{k_1}| = k_1 = \frac{2\pi}{\lambda}$,

$$q = 2k_1 \sin(\Theta) = \frac{4\pi}{\lambda} \sin(\Theta). \tag{1.16}$$

The amplitude of the scattered beam relates to the amplitude of the incoming wave via the electron radius r_0, which is the **scattering length** of a single free electron.

X-ray detectors can only detect the intensity of photons (in terms of counts per time and area, within a certain energy range), not their phase. The relation between amplitude E and intensity I is in the far-field approximation (i.e. the scattered spherical wave is approximated as a plane wave)

$$I(\boldsymbol{q}) = \frac{\epsilon_0 c}{2} \boldsymbol{E}(\boldsymbol{q})\boldsymbol{E}^*(\boldsymbol{q}) \propto E^2(\boldsymbol{q}), \tag{1.17}$$

where ϵ_0 is the dielectric constant and $\boldsymbol{E}^*$ is the complex conjugate of the amplitude $\boldsymbol{E}$. This brings us to the definition of the **differential scattering cross section** (Fig. 1.12c) $\frac{\mathrm{d}\sigma(\boldsymbol{q})}{\mathrm{d}\Omega}$:

---

[16] Following Eq. 1.15, the amplitude of a wave scattered by a proton is $\left(\frac{m_p}{m_0}\right)^2 = 3.37 \times 10^6$ times less than the amplitude of a wave scattered by an electron.

[17] For a detailed derivation of Eq. 1.15, the reader is referred to Berne and Pecora [140].

$$I_1 \mathrm{d}\sigma(\boldsymbol{q}) = I_2(\boldsymbol{q}) l^2 \,\mathrm{d}\Omega, \tag{1.18}$$

with Ω being the solid angle and l being the distance between the scatterer and the detector (detecting a wave of intensity I_2).

So far we considered scattering from a single free electron. In the context of scattering, an atom can be viewed as an assembly of electrons. For classic scattering experiments, anomalous effects (i.e. inelastic interactions due to the energy levels of the binding electrons) and multiple scattering events can be neglected. Then the scattering intensity of an atom, called the **atom form factor**, F_{atom}, is approximately proportional to its atomic number, Z:

$$F_{\mathrm{atom}}(\boldsymbol{q}) = \int_{\mathrm{atom}} \rho(\boldsymbol{r}) \exp\left(-i\boldsymbol{q}\boldsymbol{r}\right) \mathrm{d}\boldsymbol{r} \propto Z, \tag{1.19}$$

where $\rho(\boldsymbol{r})$ is the electron density distribution within the atom.[18] Equation 1.19 shows that the atom form factor is the Fourier transform of the electron density distribution. Combining Eqs. 1.15, 1.17 and 1.19, we obtain for the scattering intensity $I(\boldsymbol{q})$ of a hypothetical sample consisting of N uncorrelated atoms:

$$I(\boldsymbol{q}) \propto N {F_{\mathrm{atom}}}^2(\boldsymbol{q}). \tag{1.20}$$

1.2.3 Small-Angle X-Ray Scattering

Small angle scattering comprises scattering at angles 2Θ well below $10°$, i.e. structures in the range of 1–500 nm are analyzed, corresponding to scattering vectors $0.01\,\mathrm{nm}^{-1} < q < 5\,\mathrm{nm}^{-1}$. The low scattering angle end of this range is commonly referred to as USAXS (ultra ~), and the large-angle side at the boundary to WAXS is sometimes called MAXS (medium ~) [33, 139]. In other words, SAXS is sensitive to inhomogeneities in the scattering length density on the size scale of tens of nanometers. Thus we are not concerned about the electron density distribution within atoms, but rather about the distribution of phases or components of different electron density within the sample.

In analogy to the atom form factor, one can define particle form factors as the Fourier transform of the electron density distribution within the particulate structure. For instance, the **form factor** of a homogeneous sphere of radius R reads [139]:

$$f_{\mathrm{sphere}} = 3\frac{\sin(qR) - qR\cos(qR)}{(qR)^3}. \tag{1.21}$$

[18] Other definitions exist in literature, taking the form factor as the square of the definition in Eq. 1.19.

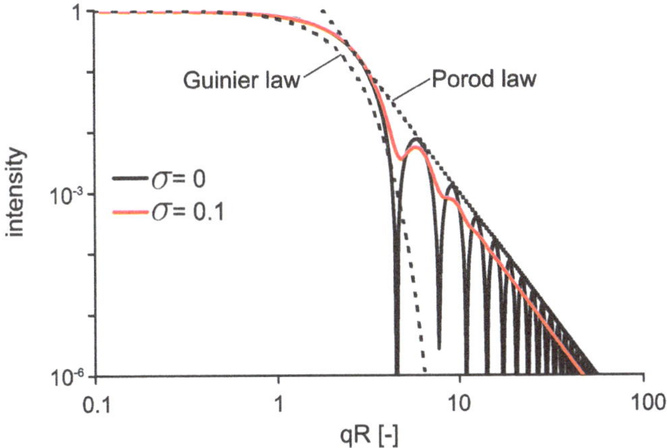

Fig. 1.13 Form factors of a monodisperse ($\sigma = 0$) and polydisperse ($\sigma = 0.1$) sets of spheres of radius R. A normal distribution with mean $qR = 1$ and standard deviation σ was assumed. The computed range is $0.05 < qR < 2.0$. With increasing polydispersity, the oscillations are blurred out. The Guinier and Porod laws are shown to hold at low and high qR, respectively

A computed scattering curve for a dilute system of spheres is shown in Fig. 1.13. For a number of other geometrically simple particles, the form factors can be computed numerically.

In a dilute solution of particles, interaction between the particles is negligible, and the scattering intensity is simply the sum of the scattering intensities of the individual particles.[19] If, however, the particles interact, i.e. the average interparticle distance is of the same order of magnitude as the particle size, crossterms arise and give way to the introduction of a **structure factor** $S(\boldsymbol{q})$:

$$I(\boldsymbol{q}) \propto V_{\mathrm{p}} F(\boldsymbol{q})^2 S(\boldsymbol{q}), \tag{1.22}$$

with V_{p} being the particle volume fraction. Consequently, for dilute systems, the structure factor $S(\boldsymbol{q})$ reduces to unity.

While the interpretation of scattering patterns in terms of form and structure factors requires an a priori model of the particle shape and interaction, model-free approaches are available. Considering that the scattering intensity is proportional to the square of the Fourier transform of the real-space electron density distribution (Eq. 1.20), one can go in the reverse direction and go from scattering space (also called s-space or reciprocal space) to real space by performing an inverse Fourier transform. Due to the phase problem, the recovery of the electron density distribution is impossible.[20] Instead, one obtains the **correlation function** $\gamma(\boldsymbol{r})$:

[19] Dispersions with a particle volume fraction below 1 % are typically considered dilute [139].

[20] Theoretically, the complete real space information can be restored under certain conditions if the coherence time of the incident beam is long relative to the exposure time. This is part of the motivation behind the development of X-ray free electron lasers (XFEL).

$$\gamma(\boldsymbol{r}) = \mathcal{F}^{-1}(I(\boldsymbol{q})). \tag{1.23}$$

One also arrives at the correlation function by inserting Eq. 1.15 into Eq. 1.17:

$$I(\boldsymbol{q}) \propto \int_{\mathrm{V}} \rho(\boldsymbol{r_1}) \exp(-i\boldsymbol{q}\boldsymbol{r_1})\, \mathrm{d}\boldsymbol{r_1} \int_{\mathrm{V}} \rho(\boldsymbol{r_2}) \exp(i\boldsymbol{q}\boldsymbol{r_2})\, \mathrm{d}\boldsymbol{r_2} \tag{1.24}$$

With $\boldsymbol{r} = \boldsymbol{r_1} - \boldsymbol{r_2}$, it follows:

$$I(\boldsymbol{q}) \propto \int_{\mathrm{V}} \int_{\mathrm{V}} \rho(\boldsymbol{r_1})\rho(\boldsymbol{r_1} - \boldsymbol{r}) \exp(-i\boldsymbol{q}\boldsymbol{r})\, \mathrm{d}\boldsymbol{r_1}\, \mathrm{d}\boldsymbol{r}, \tag{1.25}$$

where the correlation function $\gamma(\boldsymbol{r})$ can be identified as the autocorrelation of the electron density distribution:

$$\gamma(\boldsymbol{r}) = \int_{\mathrm{V}} \rho(\boldsymbol{r_1})\rho(\boldsymbol{r_1} - \boldsymbol{r})\, \mathrm{d}\boldsymbol{r_1}. \tag{1.26}$$

Instructive illustrations regarding the interpretation of the correlation function can be found e.g. in Refs. [33] and [141].

Other functions have been derived from the correlation function γ, following the principle of edge enhancement. The **interface distribution function (IDF)** is the second derivative of the one-dimensional correlation function; and the **chord distribution function (CDF)** is the three-dimensional equivalent of the IDF, mainly useful for anisotropic materials, obtained from the Laplacian of the three-dimensional correlation function [33]. These model-free approaches have certain limitations. First, to perform a Fourier transform, the experimental data is required to cover the complete reciprocal space. This of course is not feasible in practice. Scattering data for $\boldsymbol{q} \to 0$ has to be extrapolated. Merging data from different detector distances is advisable to cover a broader $\boldsymbol{q}$-range. Because a SAXS pattern only represents a slice of the reciprocal space, the reciprocal space in all three dimension can only be reconstructed under certain conditions. Either the sample fulfills fiber symmetry around an axis perpendicular to the X-ray beam, or the sample has to be rotated, which then requires a computationally expensive postprocessing of the data. The interpretation of a SAXS pattern, either directly or in real space representation, is greatly exacerbated by polydispersity and diffuse boundaries, i.e. gradual changes in electron density. Both issues are present in the systems studied in this work. The **Magic Square** gives a nice overview of the relations between real and reciprocal space (Fig. 1.14) [33].

An alternative approach is to postulate a structure in real space, based on direct interpretations of the scattering data, or based on the model-free approaches outlined above, or following other assumptions. Since the transformation from real space to reciprocal space is unequivocal, the scattering patterns can either be computed numerically or even analytically (in the case of simple geometries, cf. Eq. 1.21). Then the computed pattern is compared to the experimental one and the process

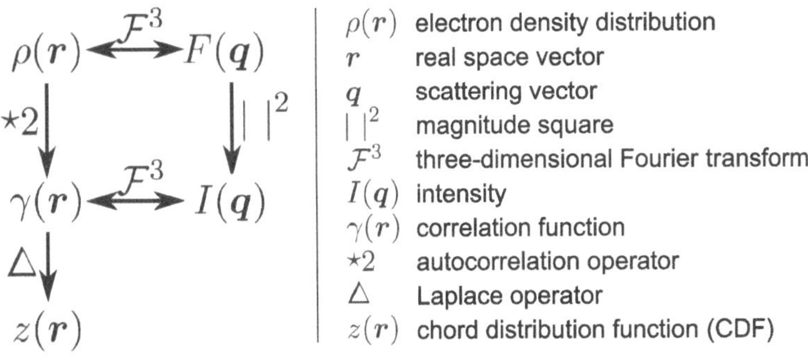

Fig. 1.14 Magic square to illustrate the relations between real space and reciprocal space functions (Adapted from Ref. [33])

is iterated until sufficient agreement is achieved. However, the obtained real-space structure is not unique.

For the special case of mass and surface fractals, the relations outlined in Sect. 1.1.3 hold.

Even though they can rarely provide a quantitative and precise evaluation of scattering data, some important very basic interpretations should be mentioned here. Historically, the Guinier law and the Porod law are of great importance. They still serve as a first rough guide in the visual interpretation of scattering curves. The **Guinier law** (Guinier 1939) holds for small q ($q \ll R_{g,e}$) and represents the outcome of the first term of a Taylor series expansion of the scattering intensity of isolated, monodisperse, spherical particles[21]:

$$I(q) \propto \exp\left(-\frac{q^2 {R_{g,e}}^2}{3}\right), \tag{1.27}$$

where $R_{g,e}$ is the electronic radius of gyration. In a Guinier plot, log (I) is plotted over q^2 and $R_{g,e}$ can be obtained from the slope of the curve. While the Guinier regime is seen as a distinct knee in a classic log (I) versus log (q) scattering curve (Fig. 1.13) and the position of the knee allows first conclusions regarding $R_{g,e}$, this feature is often blurred out by polydispersity or completely obscured due to concentration effects [142, 143].

At large q ($q \gg R_{g,e}$, large in the context of small-angle scattering, i.e. inhomogeneities in the electron density on the scale of nanometers), the **Porod law** (Porod 1951) holds:

$$I(q) \propto \frac{S_V}{q^4}, \tag{1.28}$$

[21] For a derivation, see e.g. Guinier [136].

with S_V being the specific surface. Porod's law holds for well-defined smooth surfaces. If the density changes gradually or the surface is rough, the exponent deviates from 4 (cf. Sect. 1.1.3).

As pointed out above, scattering is the result of inhomogeneities in the electron density. Not only the shape of the scattering curve deserves interpretation, but also the absolute intensity of the scattering signal allows conclusions with respect to the structure. The better the contrast between the phases (i.e. the larger the difference in electron density), the higher the **absolute scattering intensity** Q. Owing to the fact that this quantity is independent of the shape of the scatterers, it is also termed **scattering invariant**. For a two-phase system, it reads

$$Q = \int \int \int_{\boldsymbol{q}' \to 0}^{\boldsymbol{q}' \to \infty} I(\boldsymbol{q}) \, \mathrm{d}\boldsymbol{q} \propto \Phi_1 \Phi_2 (\rho_1 - \rho_2)^2 . \tag{1.29}$$

Φ_1 and Φ_2 are the volume fractions of phases 1 and 2, respectively, and ρ_1 and ρ_1 are their electron densities. The fact that Q is insensitive to the inversion of the volume fractions is called **Babinet's principle**. Since the detector plane only provides a slice through the reciprocal space, the missing third dimension has to be taken from symmetry considerations or from rotating the sample to explore the complete reciprocal space. For systems with three or more phases, the equation becomes increasingly complex [144, 145].

1.2.4 Wide-Angle X-Ray Diffraction

Wide-angle X-ray diffraction follows the same physical phenomena detailed in Sect. 1.2.1. However, the regular periodic order of scatterers gives rise to interference peaks, called diffraction.[22] The regular order on the size scale probed by WAXD ($\Theta > 10°$) corresponds to the lattice spacing d in crystalline polymers. In order to observe constructive interference between scattered waves, the phase shift between the two incoming wavelets in Fig. 1.15 must be multiples of the wavelength [147]:

$$m\lambda = 2d \sin(\Theta), \tag{1.30}$$

where m is the order of diffraction.[23] Equation 1.30 is called **Bragg's law**. It simplifies the scattering problem to a planar case. For the three-dimensional case it is instructive to think in reciprocal space. In the special case of an orthorhombic crystal structure, like in NR, the base vectors of a unit cell of the **reciprocal lattice** are collinear with the lattice unit vectors in real space. Only their lengths are inverted. In order to fulfill the diffraction condition for a given crystallographic plane, the scattering vector

[22] For an overview of diffraction theories, see e.g. Ewald [146].

[23] In polymer science, higher order diffraction peaks are frequently concealed due to imperfections and finite size effects.

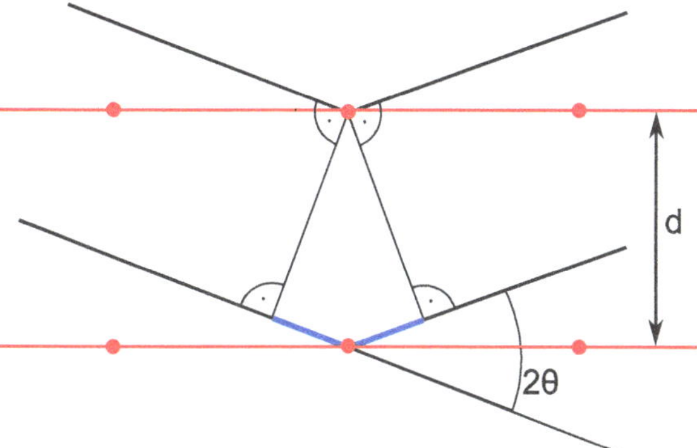

Fig. 1.15 Illustration of the diffraction condition. The *red lines* represent the crystal lattice; the *black lines* represent the X-ray photon paths. To fulfill Bragg's law, the *blue line* must be of length $m\lambda$

$\boldsymbol{q}$ must be equal to the corresponding reciprocal lattice vector. This mathematical relation is geometrically expressed in the **Ewald sphere** construction. In line with Fig. 1.15, diffraction occurs when a lattice point lies on the Ewald sphere, which is defined by its center O being at $\boldsymbol{k_1}$ and its radius $|\boldsymbol{k_1}| = \frac{2\pi}{\lambda}$ (Fig. 1.16). For a detailed derivation of the Ewald sphere construction see e.g. Ref. [148].

In numerous scattering and diffraction experiments in polymer science, advantage is taken of the **fiber symmetry** of the sample. This means that the sample structure is symmetric to a rotational axis. In particular, in a unixial tensile experiment, the symmetry axis coincides with the tensile axis, as long as the two transversal dimensions of the sample are similar. When fiber symmetry is fulfilled, we can replace the reciprocal lattice by an infinite number of lattices sharing the rotational axis, which is the c-axis in the case of NR. Or, if we stick with one lattice, the reciprocal lattice point does not need to lie on the Ewald sphere any more; instead, it is sufficient if the distance of the reciprocal lattice point from the axis of fiber symmetry equals k_1.

As opposed to metals, crystallizing polymers never become completely crystalline due to steric hindrance and due to entropy reasons [149]. Common degrees of crystallinity for semicrystalline polymers range between 10 and 70 % maximum. The remainder stays amorphous, which gives rise to the so-called amorphous halo in the wide-angle regime. The degree of crystallinity is reflected in the diffractogram as the ratio of the intensity of the crystalline diffraction peaks to the intensity of the amorphous halo [33, 150–152]. However, this methods only yields a relative number for the crystallinity, such that a normalization with independent measurements (DSC or dilatometry) is required. Alternatively, one can evaluate the decrease in the amorphous halo for itself [86, 153, 154]. An overview of methods for the determination of crystallinity is listed in Ref. [155].

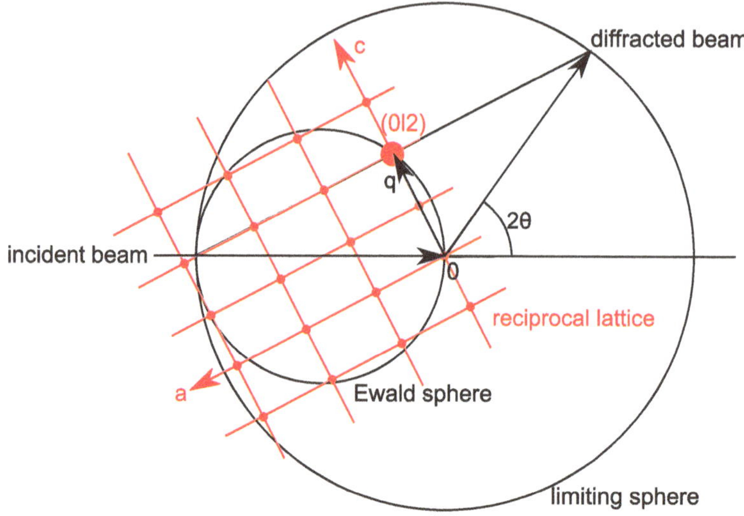

Fig. 1.16 Construction of the Ewald sphere (sphere of reflection) and the limiting sphere, exemplified for diffraction of the (0l2) crystal plane. View along the *b*-axis (Adapted from Ref. [148])

Besides the degree and orientation of the crystallites, a WAXD diffractogram also reveals information about the size of the crystallites. Following Bragg's equation, the diffraction peaks should come out as infinitely sharp point-like peaks (Dirac deltas). However, in reality the peaks are observed to have a certain width. Besides instrumental broadening, the reasons for this can be imperfections in the crystalline stacking (e.g. due to strains) or a finite size of the crystallites [148]. The **Scherrer equation** relates the width of the crystalline peak $\Delta\Theta$ to the size of the crystallite normal L_{hkl} to the scattering plane (hkl):

$$L_{hkl} = \frac{B\lambda}{\Delta\Theta \cos(\Theta)}. \tag{1.31}$$

B is a constant of the order of 1. Equation 1.31 can be readily derived from geometrical considerations regarding the limiting cases allowing constructive interference of the two crystallographic planes lying at opposite surfaces of the crystallite. In terms of Fourier space, the peak broadening can be seen as the convolution of the crystalline Dirac peak with the Fourier transform of the correlation function of the crystallites [33]. Theoretically, the contributions from finite crystallite size and lattice deformations can be separated by evaluating higher order peaks. But since higher order peaks are rarely seen in strain-crystallized NR and the crystallites are assumed to be polydisperse and in different states of deformation, a separation is almost impossible.

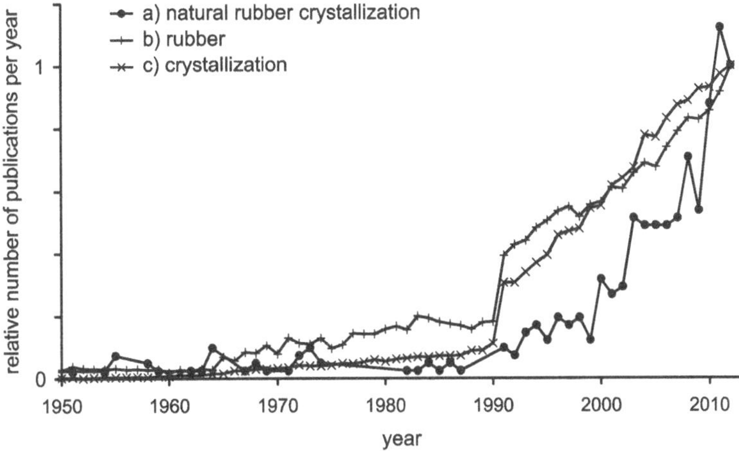

Fig. 1.17 Number of publications per year assigned to the topic of (a) *natural rubber crystallization* according to the Web of Knowledge. Given that the total number of publications per year increases continuously in almost all fields, only a careful look at the slope of the increase can identify research fields gaining importance. Thus, for comparison, the search for the keywords (b) *rubber* and (c) *crystallization* is shown. All data is normalized with respect to the 2012 figures. Total number of papers for (a) is 401, for (b) is 48347, for (c) is 128696. The rise in (a) since 2000 is very prominent and is not observed in (b) and (c). As of May 14, 2013

1.3 Crystallization in Natural Rubber

1.3.1 General Aspects of Crystallization in Natural Rubber

The earliest studies of structural changes in rubbers under strain date back to the 1920s, when Katz published his report on strain-induced crystallization[24] in natural rubber observed by WAXD [156]. Even though the fundaments of macromolecular chemistry were still lacking at that time, he concluded that highly stretched natural rubber consists of two phases and that the phase transition is reversible when the strain is removed. Since the early 2000s, the number of papers dealing with SIC has been increasing owing to the availability of more powerful synchrotron sources (Fig. 1.17). SIC attracts so much attention in the research community because it makes a big contribution to the outstanding properties of NR (Sect. 1.1.4.4). Natural rubber crystallizes in an orthorhombic[25] unit cell with lattice dimensions $a = 1.24$ nm, $b = 0.88$ nm, $c = 0.82$ nm [157–161]. A unit cell, containing two mers of cis-1,4 polyisoprene, is shown in Fig. 1.18.

Considering crystallization in natural rubber, one has to distinguish strain-induced crystallization (SIC) and thermal (quiescent) crystallization (TIC) [162]. Unde-

[24] Katz called it *fibering* what is now termed *strain-induced crystallization*.

[25] Some dispute exists in literature about the precise value of the β angle, taking values up to 93°, suggesting a monoclinic unit cell.

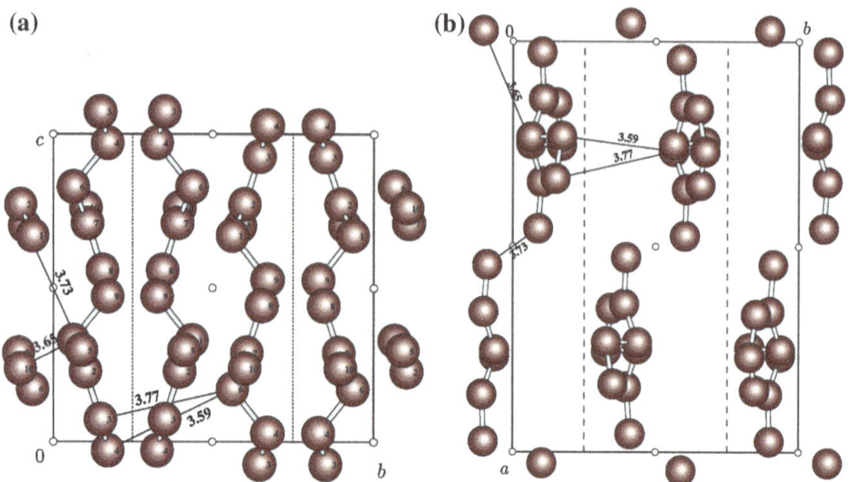

Fig. 1.18 Unit cell of crystalline natural rubber. **a** view along the *a*-axis on the *bc* plane; **b** view along the *c*-axis on the *ab* plane (Adapted with permission from Ref. [158], copyright 2005 American Chemical Society)

formed natural rubber undergoes significant thermal crystallization at temperatures below 0 °C with a maximum crystallization rate around −28 °C [16, 163]. The thermodynamic melting point lies considerably above 0 °C [80, 164]. With increasing strain, the melting temperature increases [60]. At room temperature, SIC sets in at a strain of roughly 300 % in unfilled natural rubber [165]. Above 70 °C, no SIC is observed [60]. It should be noted that, despite having the same crystal structure, the morphologies of thermally induced crystallites and strain induced crystallites are completely different [166, 167]. Quiescent crystallization forms spherulites [168, 169], whereas SIC promotes the formation of highly oriented fibrillar structures (Fig. 1.19) [167, 170]. In spherulites, folded-chain lamellae are arranged isotropically, giving rise to an unoriented WAXD pattern [171]. The periodic spacing (long period) of the lamellae can be inferred from the long period ring position in a SAXS pattern [169, 172]. The long period is in the range of 20 nm, depending on temperature [169]. In contrast to TIC, strain-crystallized NR contains semicrystalline strands, highly oriented along the stretching direction. These strands are made up of small extended chain crystallites, interrupted by amorphous segments due to entanglements and crosslinks, which cannot be transformed to the crystalline state. The strands are roughly 10–25 nm in diameter [173]. Historically, these crystallites were called γ-fibrils by Andrews, and are now generally referred to as shish [162, 174, 175]. Cooling of strain-crystallized NR can produce intermediate structures, depending on the strain and temperature. In this case, the shish serves as a backbone for lamellar overgrowth, forming folded-chain α-crystallites (or kebab) perpendicular to the backbone [174].

In the following, only SIC is considered.

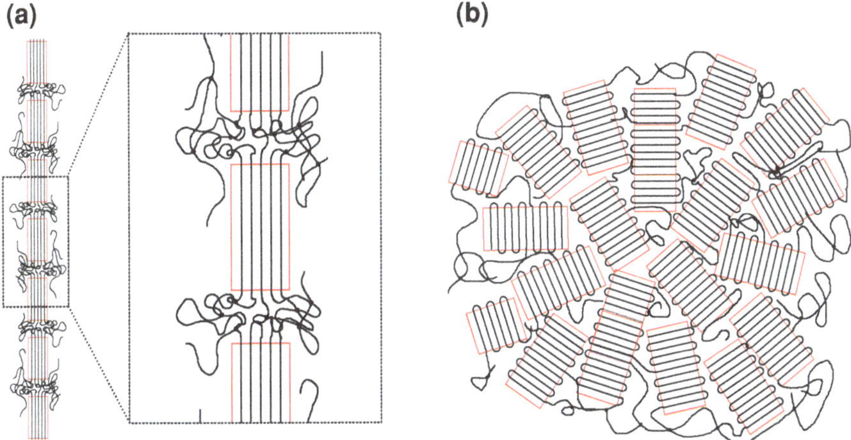

Fig. 1.19 Crystal morphologies in semicrystalline NR. **a** Shish structure resulting from strain-induced crystallization. Stretching direction is vertical. **b** Spherulitic morphology from thermal quiescent crystallization. For simplification, the chains are drawn with adjacent re-entry. In both cases, entanglements and crosslinks are restricted to the amorphous phase

While SIC can be analyzed by a range of methods, e.g. DSC [176, 177], NMR [60, 175], IR spectroscopy [175] and dilatometry [176], the most widely used method is WAXD. The relations between strain, temperature and crystallinity have been studied extensively [178].

At first sight, the fact that in unfilled rubber the SIC sets in at strains as large as 300 %, suggests that SIC might be of little relevance in real-life loading conditions of most rubber products. However, in reality the local strain often considerably exceeds the nominal strain. In filled rubbers, due to the rigidity of the filler phase, the matrix phase has to bear the complete deformation, such that the matrix strain is considerably larger than the external strain, especially when the rubber is highly filled. This effect is referred to as strain amplification (Sect. 1.1.4). Thus, in filled rubbers, the SIC onset strain can be reduced to around 150 % strain. Second, certain geometries give rise to local strain concentration, e.g. around a crack tip. Theoretically, when approaching a crack tip, one encounters a strain singularity. In reality, the local strain concentration leads to a crystalline zone around the crack tip which locally reinforces the material and thus slows down the crack growth.

1.3.2 Kinetics of Crystallization

Besides the effects of strain and temperature, the time-dependency of SIC should not be overlooked. Since most rubber products, especially tires, are subjected to dynamic loads, the structure under realistic loading conditions can considerably

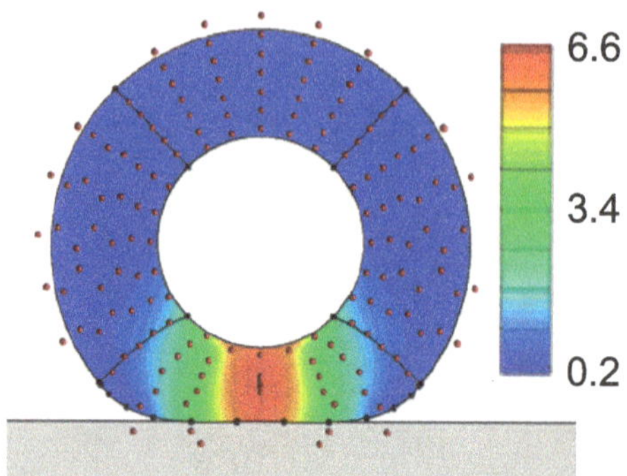

Fig. 1.20 Stress distribution in an idealized tire in contact with a rigid surface. The *color index* represents the magnitude of the first Piola-Kirchhoff stress tensor. In a rolling tire, a given material voxel experiences a pulse-like stress history (Adapted with permission from Ref. [183], copyright 2011 Elsevier)

deviate from what is suggested by quasistatic experiments (Fig. 1.20). This is due to the finite crystallization kinetics and was shown for the first time in 1932 by Acken et al., who utilized a stroboscopic technique to accumulate WAXD diffractograms over several deformation cycles to reach sufficient exposure [179]. They found that the crystallinity in the dynamically stretched rubber was considerably suppressed as compared to quasistatic experiments. This type of setup was later optimized by Kawai et al. [180, 181]. Using a variable phase shift between the stroboscope and the dynamic stretching device, the complete crystallinity versus strain curve could be reconstructed. Recently, Candau et al. extracted a crystallinity versus time curve from stroboscopic dynamic experiments [182]. They defined the characteristic time as the time period between the onset of crystallization and the time at maximum strain in a cyclic experiment. By varying the stretching frequency, they constructed a crystallinity versus characteristic time plot, which was fitted with a stretched exponential function. They obtained a characteristic crystallization time of 20 ms.

The alternative approach to cyclic loading is to apply a steplike loading and then to follow the crystallization over time. The pioneering work in this respect was done by Dunning and Pennells in 1967, who took advantage of a continuous steplike strain when a rubber band is passed over two rolls rotating at different speeds [184]. They reported incubation times between 5 ms and 18 s, depending on the strain level. However, the fact that the initial scattered intensity in the region of the diffraction spots increases with strain, suggests that crystallization might have set in even at times shorter than the experimental time scale.

Mitchell and Meier performed experiments on a high-speed tensile tester. They utilized the enthalpy of crystallization and employed thermal techniques to measure the crystallinity after a steplike loading [185]. This method relies on several assumptions in order to separate crystallization enthalpy from the work of deformation. The crystallinity versus time behavior was fitted with a first order rate law with a half time of 45 ms.

Only in the last few years, owing to the availability of more powerful synchrotrons, did the direct measurement of crystallinity after a step strain become possible. Tosaka et al. performed WAXD experiments with a pattern acquisition rate of approximately

12 Hz [8, 186, 187]. Despite relatively slow steplike loading (usually $10\,s^{-1}$), they reported that the crystallization process only begins after the loading step is finished, following a two-step kinetic law. After roughly 10 s the crystallinity was found to approach a steady state.

To date, the understanding of the molecular processes and mechanism behind the crystallization kinetics is still very limited. An interpretation of strain-induced crystallization on the molecular level, e.g. in analogy to the well-known nucleation and growth models for quiescent crystallization, is still lacking [188]. In that sense, the SIC theory lags behind the TIC theory by several decades. De Gennes put forward the idea of an instantaneous coil-stretch transition [189]. Later Hsiao et al. proposed to apply this transition to chain segments between crosslinks [190], considering that the crystallites are much smaller than the typical distance between crosslinks or entanglements, which is of the order of 20–150 nm [10, 191, 192].

In the context of thermal crystallization, the most famous kinetic equation is the Avrami equation [193]. It has been modified and extended since its introduction in the 1940s. However, it only considers the superstructure of the crystallites on the level of the ordering of lamellae into spherulites and the growth of the spherulites. It does not address the transition from the random coil state to lamellae. Therefore, it is not applicable to SIC.

1.4 Cavitation

It is commonly accepted that the presence of cavities[26] in deformed rubber interferes with the mechanical behavior of the material. On the one hand, the energy dissipation involved in the formation of cavities toughens the material. On the other hand, cavities can impair the mechanical properties and the growth of cavities is considered to play a major role in the early stages of crack propagation [101, 194–197]. Due to its long history and due to the importance of the subject, it has been reviewed several times [194, 198, 199].

The most widely used method to detect cavitation is dilatometry during deformation [163, 176, 200–203]. It was pointed out that due to the diffusion properties of rubber, gas dilatometers cannot be used to determine cavitation [176]. Also the superposition of volume changes by cavitation and strain-induced crystallization has to be accounted for. SIC leads to a decrease in volume by up to 2.7 % [163]. Another popular method is scattering [144, 145, 204, 205], either from X-rays or light. Due to the large difference in electron density between the cavities and the surrounding matrix, cavitation is reflected in a strong increase in total scattering intensity. The presence of elongated cavities is observed as a streak in the scattering pattern. The disadvantage of scattering methods is that they can only detect cavities within a certain narrow size range. Recently, optical methods gained popularity as a method

[26] The terms *cavitation* and *void formation* are often used interchangeably, however some authors prefer to use *voids* to describe stable hollow regions in thermoplastics and other glassy materials, whereas *cavities* is used in the context of elastomers for hollow regions growing in an unstable manner [194].

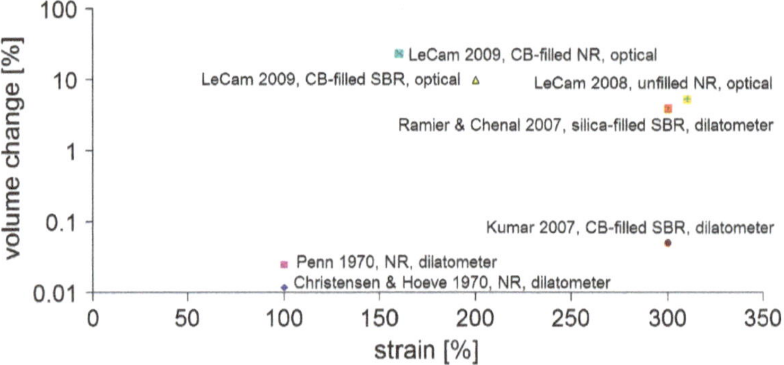

Fig. 1.21 Literature overview of volume changes in elastomers under strain, measured by various methods and on different materials. Le Cam 2008 [208], Le Cam 2009 [207], Ramier and Chenal 2007 [205, 213], Kumar 2007 [218], Penn 1970 [201], Christensen and Hoeve 1970 [200]

for volume measurement [206–208]. Furthermore, cavitation was studied by SEM [101, 209], tomography [210], acoustic emission analysis [211] and NMR [197].

Despite the multitude of investigations, no clear picture about the quantity of cavitation could be obtained so far (Fig. 1.21). However, most reports agree that in unfilled rubber, cavitation is negligible [200–202, 205, 212]. Few authors reported a significant volume increase in unfilled rubber [207, 208, 213], which might be ascribed to the high zinc oxide content in the *unfilled*[27] samples under study. For filled rubbers, most studies reported volume increases between 1 and 5 % [176, 200, 202, 205, 213], whereas LeCam found a volume increase by as much as 25 %.[28] The results suggest that cavitation is more prominent when filler loadings are high, the adhesion between filler and matrix is low and the strain is large [205, 209, 214].

The mechanical criteria for cavitation are still subject to discussion. The most famous approach is due to Gent (1958), who argued that the critical stress is $\frac{5}{6}E$, with E being the tensile modulus [215]. Despite the good agreement with experimental results, he later noted that this simple relation only holds for a certain limited particle size range [144, 199, 216, 217]. A different approach follows the Griffith criterion in the derivation of a cavitation criterion [194]. It supports experimental evidence that a certain minimum initial flaw size is required in order to expand the flaw to a cavity. It is commonly assumed that defects of a size larger than 100 nm exist in any rubber material, such that the cavitation problem can be reduced to the growth of these defects.

[27] *Unfilled* typically specifies materials without reinforcing filler. Recipes containing additives like zinc oxide are thus classified as *unfilled*.

[28] It should be noted that the rubbers studied by LeCam contained the unusually high amount of 10 phr of zinc oxide, which possibly stimulates cavitation.

References

1. J. Loadman, *Tears of the Tree: The Story of Rubber—A Modern Marvel* (Oxford University Press, Oxford, 2005)
2. A. Hwee, Y. Tanaka, Structure of natural rubber. Trends Polym. Sci. (Oxford) **3**(5), 493–513 (1993)
3. Y. Tanaka, Structural characterization of natural polyisoprenes: solve the mystery of natural rubber based on structural study. Rubber Chem. Technol. **74**(3), 355–375 (2001)
4. S. Toki et al., Entanglements and networks to strain-induced crystallization and stress-strain relations in natural rubber and synthetic polyisoprene at various temperatures. Macromolecules **46**(13), 5238–5248 (2013)
5. S. Toki et al., New insights into the relationship between network structure and strain-induced crystallization in un-vulcanized and vulcanized natural rubber by synchrotron X-ray diffraction. Polymer **50**(9), 2142–2148 (2009)
6. A.N. Gent, *Engineering with Rubber: How to Design Rubber Components* (Hanser Gardner, Cincinnati, 2001)
7. S. Trabelsi, P.A. Albouy, J. Rault, Stress-induced crystallization properties of natural and synthetic cis-polyisoprene. Rubber Chem. Technol. **77**(2), 303–316 (2004)
8. M. Tosaka et al., Crystallization of stretched network chains in cross-linked natural rubber. J. Appl. Phys. **101**, 084909 (2007)
9. Y. Miyamoto, H. Yamao, K. Sekimoto, Crystallization and melting of polyisoprene rubber under uniaxial deformation. Macromolecules **36**(17), 6462–6471 (2003)
10. M. Tosaka, Strain-induced crystallization of crosslinked natural rubber as revealed by X-ray diffraction using synchrotron radiation. Polym. J. **39**(12), 1207–1220 (2007)
11. G.G. Odian, *Principles of Polymerization* (Wiley, New York, 2007)
12. K. Nagdi, *Rubber as an Engineering Material: Guideline for Users* (Oxford University Press, New York, 1993)
13. G. Heideman et al., Activators in accelerated sulfur vulcanization. Rubber Chem. Technol. **77**(3), 512–541 (2004)
14. R. Steudel, Y. Steudel, Interaction of zinc oxide clusters with molecules related to the sulfur vulcanization of polyolefins (rubber). Chem. Eur. J. **12**(33), 8589–8602 (2006). ISSN: 1521-3765
15. P. Ghosh et al., Sulfur vulcanization of natural rubber for benzothiazole accelerated formulations: from reaction mechanisms to a rational kinetic model. Rubber Chem. Technol. **76**(3), 592–693 (2003). ISSN: 0035-9475
16. L.R.G. Treloar, *The Physics of Rubber Elasticity*, 3rd edn., Oxford Classic Texts in the Physical Sciences (Clarendon Press, Oxford, 2009)
17. J.S. Dick, *Rubber Technology: Compounding and Testing for Performance* (Hanser Gardner, Cincinnati, 2009). ISBN: 156990278X
18. B. Rodgers et al., *Carbon Black* (Marcel Dekker, New York, 2004)
19. M. Klüppel, A. Schröder, G. Heinrich, Carbon black, in *Physical Properties of Polymers Handbook*, ed. by M.E. James (Springer, New York, 2007). ISBN: 0387312358
20. M. Klüppel, G. Heinrich, Fractal structures in carbon black reinforced rubbers. Rubber Chem. Technol. **68**, 623 (1995)
21. G.J. Schneider et al., Correlation of mass fractal dimension and cluster size of silica in styrene butadiene rubber composites. J. Chem. Phys. **133**, 094902 (2010)
22. A.I. Medalia, F.A. Heckman, Morphology of aggregates II. Size and shape factors of carbon black aggregates from electron microscopy. Carbon **7**(5), 567–582 (1969)
23. T. Koga et al., New insight into hierarchical structures of carbon black dispersed in polymer matrices: a combined small-angle scattering study. Macromolecules **41**(2), 453–464 (2008)
24. J. Hyeon-Lee et al., Fractal analysis of flame-synthesized nanostructured silica and titania powders using small-angle X-ray scattering. Langmuir **14**(20), 5751–5756 (1998)

25. G.J. Schneider, D. Göritz, A novel model for the interpretation of small-angle scattering experiments of self-affine structures. J. Appl. Crystallogr. **43**(1), 12–16 (2009). ISSN: 0021-8898
26. D.J. Kohls, G. Beaucage, Rational design of reinforced rubber. Curr. Opin. Solid State Mater. Sci. **6**(3), 183–194 (2002). ISSN: 1359-0286
27. G.J. Schneider, Analyse der Struktur von aktiven Füllstoffen mittels Streumethoden. Ph.D. Thesis, Universität Regensburg (2006)
28. B.B. Mandelbrot, *The Fractal Geometry of Nature* (Times Books, New York, 1983)
29. M. Fleischmann, D.J. Tildesley, R.C. Ball, *Fractals in the Natural Sciences: A Discussion* (Princeton University Press, Princeton, 1990). ISBN: 0691085617
30. G. Huber, T.A. Vilgis, On the mechanism of hydrodynamic reinforcement in elastic composites. Macromolecules **35**(24), 9204–9210 (2002)
31. H.D. Bale, P.W. Schmidt, Small-angle X-ray-scattering investigation of submicroscopic porosity with fractal properties. Phys. Rev. Lett. **53**(6), 596–599 (1984)
32. P.W. Schmidt, Small-angle scattering studies of disordered, porous and fractal systems. J. Appl. Crystallogr. **24**(5), 414–435 (1991)
33. N. Stribeck, *X-Ray Scattering of Soft Matter* (Springer, Heidelberg, 2007)
34. J.E. Martin, Scattering exponents for polydisperse surface and mass fractals. J. Appl. Crystallogr. **19**(1), 25–27 (1986). ISSN: 0021-8898
35. G. Beaucage et al., Morphology of polyethylene-carbon black composites. J. Polym. Sci. B Polym. Phys. **37**(11), 1105–1119 (1999). ISSN: 1099-0488
36. R. Hjelm et al., The microstructure and morphology of carbon black: a study using small angle neutron scattering and contrast variation. J. Mater. Res. **9**(12), 3210–3222 (1994)
37. W. Ruland, Apparent fractal dimensions obtained from small-angle scattering of carbon materials. Carbon **39**(2), 323–324 (2001)
38. G. Heinrich, M. Klüppel, A hypothetical mechanism of carbon black formation based on molecular ballistic deposition. Kautschuk, Gummi, Kunststoffe **54**(4), 159–165 (2001)
39. G. Beaucage, Approximations leading to a unified exponential/power-law approach to small-angle scattering. J. Appl. Crystallogr. **28**(6), 717–728 (1995)
40. M. Klüppel, The role of disorder in filler reinforcement of elastomers on various length scales. Adv. Polym. Sci. **164**, 1–86 (2003)
41. A. Schröder, M. Klüppel, R. Schuster, Characterisation of surface activity of carbon black and its relation to polymer-filler interaction. Macromol. Mater. Eng. **292**(8), 885–916 (2007)
42. T.P. Rieker, S. Misono, F. Ehrburger-Dolle, Small-angle X-ray scattering from carbon blacks: crossover between the fractal and Porod regimes. Langmuir **15**(4), 914–917 (1999)
43. T.P. Rieker, M. Hindermann-Bischoff, F. Ehrburger-Dolle, Small-angle X-ray scattering study of the morphology of carbon black mass fractal aggregates in polymeric composites. Langmuir **16**(13), 5588–5592 (2000)
44. J. Fröhlich, S. Kreitmeier, D. Göritz, Surface characterization of carbon blacks. Kautsch. Gummi Kunstst. **51**, 370–376 (1998)
45. C.R. Herd, G.C. McDonald, W.M. Hess, Morphology of carbon-black aggregates: fractal versus Euclidean geometry. Rubber Chem. Technol. **65**, 107 (1992)
46. R. Botet, R. Jullien, Fractal aggregates of particles. Phase Trans. Multinat. J. **24**(2), 691–736 (1990)
47. T.C. Gruber, T.W. Zerda, M. Gerspacher, 3D morphological characterization of carbon-black aggregates using transmission electron microscopy. Rubber Chem. Technol. **67**(2), 280–287 (1994)
48. J. Feder, *Fractals* (Plenum Press, New York, 1988)
49. K.J. Falconer, *Fractal Geometry* (Wiley, New York, 1990)
50. G. Heinrich, M. Klüppel, T. Vilgis, Reinforcement theories, in *Physical Properties Of Polymers Handbook*, ed. by M.E. James (Springer, New York, 2007). ISBN: 0387312358
51. D.C. Edwards, Polymer-filler interactions in rubber reinforcement. J. Mater. Sci. **25**(10), 4175–4185 (1990)

52. T.A. Vilgis, G. Heinrich, Disorder-induced enhancement of polymer adsorption—a model for the rubber-polymer interaction in filled rubbers. Macromolecules **27**(26), 7846–7854 (1994)
53. A.R. Payne, Dynamic properties of heat-treated butyl vulcanizates. J. Appl. Polym. Sci. **7**, 873–885 (1963)
54. S. Merabia, P. Sotta, D.R. Long, A microscopic model for the reinforcement and the nonlinear behavior of filled elastomers and thermoplastic elastomers (Payne and Mullins effects). Macromolecules **41**(21), 8252–8266 (2008)
55. S.D. Gehman, J.E. Field, X-ray structure of rubber-carbon black mixtures. Ind. Eng. Chem. **32**(10), 1401–1407 (1940)
56. J. Domurath et al., Modelling of stress and strain amplification effects in filled polymer melts. J. Non-Newton. Fluid Mech. **171**, 8–16 (2012)
57. L. Mullins, N.R. Tobin, Stress softening in rubber vulcanizates. I. Use of a strain-amplification factor to describe the elastic behavior of filler-reinforced vulcanized rubber. J. Appl. Polym. Sci **9**, 2993 (1965)
58. A.I. Medalia, Morphology of aggregates: I. Calculation of shape and bulkiness factors; application to computer-simulated random flows. J. Colloid Interface Sci. **24**(3), 393–404 (1967)
59. S. Westermann et al., Strain amplification effects in polymer networks. Physica B: Cond. Matter **234**, 306–307 (1997). ISSN: 0921-4526
60. J. Rault et al., Stress-induced crystallization and reinforcement in filled natural rubbers: 2H NMR study. Macromolecules **39**(24), 8356–8368 (2006)
61. G. Huber, T.A. Vilgis, Universal properties of filled rubbers: mechanisms for reinforcement on different length scales. Kautsch. Gummi Kunstst. **52**(2), 102–107 (1999)
62. D.J. Lee, J.A. Donovan, Microstructural changes in the crack tip region of carbon-black-filled natural rubber. Rubber Chem. Technol. **60**(5), 910–923 (1987)
63. J.A.C. Harwood, L. Mullins, A.R. Payne, Stress softening in natural rubber vulcanizates. Part II. Stress softening effects in pure gum and filler loaded rubbers. J. Appl. Polym. Sci. **9**(9), 3011–3021 (1965)
64. S. Dupres et al., Local deformation in carbon black-filled polyisoprene rubbers studied by NMR and X-ray diffraction. Macromolecules **42**(7), 2634–2644 (2009)
65. S. Westermann et al., Matrix chain deformation in reinforced networks: a SANS approach. Macromolecules **32**(18), 5793–5802 (1999). ISSN: 0024-9297
66. A. Botti et al., A microscopic look at the reinforcement of silica-filled rubbers. J. Chem. Phys. **124**, 174908 (2006)
67. N. Jouault et al., Direct small-angle-neutron-scattering observation of stretched chain conformation in nanocomposites: more insight on polymer contributions in mechanical reinforcement. Phys. Rev. E **82**(3), 31801 (2010). ISSN: 1063-651X
68. S. Trabelsi, P.A. Albouy, J. Rault, Effective local deformation in stretched filled rubber. Macromolecules **36**(24), 9093–9099 (2003)
69. K. Brüning, K. Schneider, G. Heinrich, Deformation and orientation in filled rubbers on the nano- and microscale studied by X-ray scattering. J. Polym. Sci. B Polym. Phys. **50**(24), 1728–1732 (2012)
70. T. Witten, M. Rubinstein, R. Colby, Reinforcement of rubber by fractal aggregates. J. Phys. II **3**(3), 367–383 (1993)
71. J. Fröhlich, W. Niedermeier, H.-D. Luginsland, The effect of filler-filler and filler-elastomer interaction on rubber reinforcement. Compos. A Appl. Sci. Manuf. **36**(4), 449–460 2005. ISSN: 1359-835X. doi:10.1016/j.compositesa.2004.10.004, http://www.sciencedirect.com/science/article/B6TWN-4F0854M-5/2/e872fc26f9297bbc0c83c9e362647db3
72. T. Lanzl, Charakterisierung von Ruß-Kautschuk-Mischungen mittels dielektrischer Spektroskopie. Ph.D. Thesis, Universität Regensburg (2001)
73. V.M. Litvinov, P.A.M. Steeman, EPDM-carbon black interactions and the reinforcement mechanisms, as studied by low-resolution 1H NMR. Macromolecules **32**(25), 8476–8490 (1999)
74. S. Wolff, M.J. Wang, E.H. Tan, Filler-elastomer interactions. Part VII. Study on bound rubber. Rubber Chem. Technol. **66**(2), 163–177 (1993)

75. S. Kohjiya, A. Kato, Y. Ikeda, Visualization of nanostructure of soft matter by 3D-TEM: nanoparticles in a natural rubber matrix. Progr. Polym. Sci. **33**(10), 979–997 (2008)
76. S. Kohjiya et al., Visualisation of carbon black networks in rubbery matrix by skeletonisation of 3D-TEM image. Polymer **47**(10), 3298–3301 (2006)
77. M.-J. Wang, Effect of polymer-filler and filler-filler interactions on dynamic properties of filled vulcanizates. Rubber Chem. Technol. **71**(3), 520–589 (1998)
78. J.T. Bauman, *Fatigue, Stress, and Strain of Rubber Components: Guide for Design Engineers* (Hanser, Cincinnati, 2008)
79. S. Toki et al., Multi-scaled microstructures in natural rubber characterized by synchrotron X-ray scattering and optical microscopy. J. Polym. Sci. A Polym. Chem. **46**(22), 2456–2464 (2008)
80. W.R. Krigbaum et al., Effect of strain on the thermodynamic melting temperature of polymers. J. Polym. Sci. A-2 Polym. Phys. **4**(3), 475–489 (1966)
81. W.R. Krigbaum, R.J. Roe, Diffraction study of crystallite orientation in a stretched polychloroprene vulcanizate. J. Polym. Sci. A General Papers **2**(10), 4391–4414 (1964)
82. M. Tsuji, T. Shimizu, S. Kohjiya, TEM studies on thin films of natural rubber and polychloroprene crystallized under molecular orientation II. Highly prestretched thin films. Polym. J. **32**(6), 505–512 (2000)
83. M. Tsuji, T. Shimizu, S. Kohjiya, TEM studies on thin films of natural rubber and polychloroprene crystallized under molecular orientation. Polym. J. **31**(9), 784–789 (1999)
84. S. Toki et al., Structural developments in synthetic rubbers during uniaxial deformation by in situ synchrotron X-ray diffraction. J. Polym. Sci. B Polym. Phys. **42**(6), 956–964 (2004)
85. S. Toki et al., Strain-induced molecular orientation and crystallization in natural and synthetic rubbers under uniaxial deformation by in-situ synchrotron X-ray study. Rubber Chem. Technol. **77**, 317 (2004)
86. G.R. Mitchell, A wide-angle X-ray study of the development of molecular orientation in crosslinked natural rubber. Polymer **25**(11), 1562–1572 (1984)
87. M. Kroon, A constitutive model for strain-crystallising rubber-like materials. Mech. Mater. **42**(9), 873–885 (2010). ISSN: 0167-6636. doi:10.1016/j.mechmat.2010.07.008, http://www.sciencedirect.com/science/article/pii/S0167663610000980
88. W.V. Mars, A. Fatemi, A phenomenological model for the effect of R ratio on fatigue of strain crystallizing rubbers. Rubber Chem. Technol. **76**(5), 1241–1258 (2003)
89. R.S. Stein, J. Powers, *Topics in Polymer Physics* (Imperial College Press, Singapore, 2006)
90. L.H. Sperling, *Introduction to Physical Polymer Science* (Wiley, New York, 1992)
91. G. Marckmann, E. Verron, Comparison of hyperelastic models for rubber-like materials. Rubber Chem. Technol. **79**(5), 835–858 (2006)
92. L. Mullins, N.R. Tobin, Theoretical model for the elastic behavior of filler-reinforced vulcanized rubbers. Rubber Chem. Technol. **30**(2), 555–571 (1957)
93. F. Bueche, Mullins effect and rubber-filler interaction. J. Appl. Polym. Sci. **5**(15), 271–281 (1961)
94. G. Heinrich, M. Klüppel, Recent advances in the theory of filler networking in elastomers. Adv. Polym. Sci. **160**(1), 1–44 (2002)
95. S. Richter et al., Jamming in filled polymer systems. Macromol. Symp. **291**(1), 193–201 (2010). ISSN: 1521-3900
96. S.M. Cadwell et al., Dynamic fatigue life of rubber. Ind. Eng. Chem. Anal. Ed. **12**(1), 19–23 (1940)
97. W.V. Mars, A. Fatemi, A literature survey on fatigue analysis approaches for rubber. Int. J. Fatigue **24**(9), 949–961 (2002)
98. A. Andriyana, N. Saintier, E. Verron, Multiaxial fatigue life prediction of rubber using configurational mechanics and critical plane approaches: a comparative study, in *Proceedings of the VI. European Conference on Constitutive Models for Rubber*, vol. 5, ed. by P.E. Austrell, L. Kari (Balkema, Tokyo, 2008), p. 191
99. W.V. Mars, A. Fatemi, Factors that affect the fatigue life of rubber: a literature survey. Rubber Chem. Technol. **77**, 391–412 (2004)

100. W.V. Mars, A. Fatemi, Nucleation and growth of small fatigue cracks in filled natural rubber under multiaxial loading. J. Mater. Sci. **41**(22), 7324–7332 (2006)
101. J.B. Le Cam et al., Mechanism of fatigue crack growth in carbon black filled natural rubber. Macromolecules **37**(13), 5011–5017 (2004)
102. A.G. Thomas, The development of fracture mechanics for elastomers. Rubber Chem. Technol. **67**(3), 50–67 (1994)
103. R.S. Rivlin, A.G. Thomas, Rupture of rubber. I. Characteristic energy for tearing. J. Polym. Sci. **10**(3), 291–318 (1953)
104. W.V. Mars, A. Fatemi, Fatigue crack nucleation and growth in filled natural rubber. Fatigue Fract. Eng. Mater. Struct. **26**(9), 779–789 (2003)
105. A.G. Thomas, Rupture of rubber. II. The strain concentration at an incision. J. Polym. Sci. **18**(88), 177–188 (1955)
106. G. Andreini et al., Crack growth behavior of styrene-butadiene rubber, natural rubber, and polybutadiene rubber compounds: comparison of pure-shear versus strip tensile test. Rubber Chem. Technol. **86**(1), 132–145 (2013)
107. J.B. Le Cam, E. Toussaint, The mechanism of fatigue crack growth in rubbers under severe loading: the effect of stress-induced crystallization. Macromolecules **43**(10), 4708–4714 (2010)
108. J.R. Rice, A path independent integral and the approximate analysis of strain concentration by notches and cracks. J. Appl. Mech. **35**, 379–386 (1968)
109. G. Cherepanov, Crack propagation in continuous media. J. Appl. Math. Mech. **31**, 503–512 (1967)
110. N. Ait Hocine, M. Nait Abdelaziz, Fracture analysis of rubber-like materials using global and local approaches: initiation and propagation direction of a crack. Polym. Eng. Sci. **49**(6), 1076–1088 (2009)
111. T. Horst, Spezifische Ansätze zur bruchmechanischen Charakterisierung von Elastomeren. Ph.D. Thesis, Universität Dresden (2010)
112. K.H. Schwalbe, Approximate calculation of fatigue crack growth. Int. J. Fract. **9**(4), 381–395 (1973)
113. G. Andreini et al., Comparison of sine versus pulse waveform effects on fatigue crack growth behavior of NR, SBR, and BR compounds. Rubber Chem. Technol. **83**(4), 391–403 (2010)
114. U.G. Eisele, S.A. Kelbch, A.J.M. Summer, Crack growth performance of tire compounds. Rubber World **212**, 38–45 (1995)
115. R.J. Harbour, A. Fatemi, W.V. Mars, The effect of a dwell period on fatigue crack growth rates in filled SBR and NR. Rubber Chem. Technol. **80**(5), 838–853 (2007)
116. E.H. Andrews, Crack propagation in a strain-crystallizing elastomer. J. Appl. Phys. **32**(3), 542–548 (1961)
117. R.J. Harbour, A. Fatemi, W.V. Mars, Fatigue crack growth of filled rubber under constant and variable amplitude loading conditions. Fatigue Fract. Eng. Mater. Struct. **30**(7), 640–652 (2007)
118. A.T. Zehnder, *Fracture Mechanics* (Springer, New York, 2012)
119. N. Saintier, G. Cailletaud, R. Piques, Cyclic loadings and crystallization of natural rubber: an explanation of fatigue crack propagation reinforcement under a positive loading ratio. Mater. Sci. Eng. A **528**(3), 1078–1086 (2011). ISSN: 0921-5093. doi:10.1016/j.msea.2010.09.079, http://www.sciencedirect.com/science/article/pii/S0921509310011251
120. W.F. Busse, Tear resistance and structure of rubber. Ind. Eng. Chem. **26**(11), 1194–1199 (1934)
121. W.V. Mars, Computed dependence of rubber's fatigue behavior on strain crystallization. Rubber Chem. Technol. **82**, 51–61 (2009)
122. N. Andre, G. Cailletaud, R. Piques, Haigh diagram for fatigue crack initiation prediction of natural rubber components. Kautsch. Gummi Kunstst. **52**(2), 120–123 (1999)
123. H.W. Greensmith, Rupture of rubber. IV. Tear properties of vulcanizates containing carbon black. J. Polym. Sci. **21**(98), 175–187 (1956)

124. K. Sakulkaew, A.G. Thomas, J.J.C. Busfield, The effect of the rate of strain on tearing in rubber. Polym. Testing **30**(2), 163–172 (2011)
125. B. Gabrielle et al., Tear rotation in reinforced natural rubber, in *Proceedings of the VII. European Conference on Constitutive Models for Rubber* (2011), pp. 221–225
126. B. Gabrielle et al., Effect of tear rotation on ultimate strength in reinforced natural rubber. Macromolecules **44**(17), 7006–7015 (2011)
127. J.B. Brunac, O. Gérardin, J.B. Leblond, On the heuristic extension of Haigh's diagram for the fatigue of elastomers to arbitrary loadings. Int. J. Fatigue **31**(5), 859–867 (2009)
128. W.V. Mars, A. Fatemi, Multiaxial fatigue of rubber: Part I: Equivalence criteria and theoretical aspects. Fatigue Fract. Eng. Mater. Struct. **28**(6), 515–522 (2005)
129. W.V. Mars, A. Fatemi, Multiaxial stress effects on fatigue behavior of filled natural rubber. Int. J. Fatigue **28**(5-6), 521–529 (2006)
130. S.V. Hainsworth, An environmental scanning electron microscopy investigation of fatigue crack initiation and propagation in elastomers. Polym. Testing **26**(1), 60–70 (2007)
131. D.K. Setua, S.K. De, Scanning electron microscopy studies on mechanism of tear fracture of styrene-butadiene rubber. J. Mater. Sci. **18**(3), 847–852 (1983)
132. K. Legorju-Jago, C. Bathias, Fatigue initiation and propagation in natural and synthetic rubbers. Int. J. Fatigue **24**(2-4), 85–92 (2002)
133. D.G. Young, Dynamic property and fatigue crack propagation research on tire sidewall and model compounds. Rubber Chem. Technol. **58**(4), 785–805 (1985)
134. DESY website, http://photon-science.desy.de/facilities/petra_iii/index_eng.html. Accessed 16 May 2013
135. Lecture Notes "Physik mit Synchrotronstrahlung", given by Prof. C. Schroer, TU Dresden (2011)
136. A. Guinier et al., *Small-Angle Scattering of X-Rays* (Wiley, New York, 1955)
137. L.A. Feigin, D.I. Svergun, *Structure Analysis by Small-Angle X-Ray and Neutron Scattering* (Plenum Press, New York, 1987)
138. O. Glatter, O. Kratky, *Small Angle X-Ray Scattering* (Academic Press, London, 1982)
139. P. Lindner, T. Zemb, *Neutrons, X-Rays, and Light: Scattering Methods Applied to Soft Condensed Matter* (Elsevier North-Holland, Boston, 2002)
140. B.J. Berne, R. Pecora, *Dynamic Light Scattering, with Applications to Chemistry, Biology, and Physics* (Dover, New York, 2000)
141. C.G. Vonk, G. Kortleve, X-ray small-angle scattering of bulk polyethylene. Colloid Polym. Sci. **220**(1), 19–24 (1967)
142. G. Beaucage, H.K. Kammler, S.E. Pratsinis, Particle size distributions from small-angle scattering using global scattering functions. J. Appl. Crystallogr. **37**(4), 523–535 (2004)
143. J. Fröhlich, Analyse der räumlichen Struktur und der Oberfläche von aktiven Füllstoffen mittels Streumethoden. Ph.D. Thesis, Universität Regensburg (1998)
144. H. Zhang et al., Nanocavitation in carbon black filled styrene-butadiene rubber under tension detected by real time small angle X-ray scattering. Macromolecules **45**(3), 1529–1543 (2012)
145. A. Jánosi, Röntgenkleinwinkelstreuung von Mehrphasensystemen, interpretation der Elektronendichtefluktuation: II. Dreiphasensysteme. Monatsh. Chem. Chem. Mon. **114**(4), 383–387 (1983)
146. P.P. Ewald, Introduction to the dynamical theory of X-ray diffraction. Acta Crystallogr. A Cryst. Phys. Diffr. Theor. General Crystallogr. **25**(1), 103–108 (1969)
147. W.H. Bragg, W.L. Bragg, The reflection of X-rays by crystals. Proc. R. Soc. Lond. Ser. A **88**(605), 428–438 (1913)
148. L.E. Alexander, *X-Ray Diffraction Methods in Polymer Science* (Wiley-Interscience, New York, 1969)
149. G.K. Elyashevich, Thermodynamic and kinetics of orientational crystallization of flexible-chain polymers. Adv. Polym. Sci. **43**, 205–245 (1982). ISSN: 0065-3195
150. S. Rabiej, A comparison of two X-ray diffraction procedures for crystallinity determination. Eur. Polym. J. **27**(9), 947–954 (1991)

151. M. Tosaka et al., Orientation and crystallization of natural rubber network as revealed by WAXD using synchrotron radiation. Macromolecules **37**(9), 3299–3309 (2004)
152. J.M. Goppel, On the degree of crystallinity in natural rubber. Appl. Sci. Res. **1**(1), 3–17 (1949)
153. S.C. Nyburg, X-ray determination of crystallinity in deformed natural rubber. Br. J. Appl. Phys. **5**, 321–324 (1954)
154. R. Oono, K. Miyasaka, K. Ishikawa, Crystallization kinetics of biaxially stretched natural rubber. J. Polym. Sci. Polym. Phys. Ed. **11**(8), 1477–1488 (1973)
155. P. Bansal et al., Multivariate statistical analysis of X-ray data from cellulose: a new method to determine degree of crystallinity and predict hydrolysis rates. Bioresour. Technol. **101**(12), 4461–4471 (2010)
156. J.R. Katz, Röntgenspektrographische Untersuchungen am gedehnten Kautschuk und ihre mögliche Bedeutung für das Problem der Dehnungseigenschaften dieser Substanz. Naturwissenschaften **13**(19), 410–416 (1925)
157. C.W. Bunn, Molecular structure and rubber-like elasticity. I. The crystal structures of gutta-percha, rubber and polychloroprene. Proc. R. Soc. Lond. Ser. A. Math. Phys. Sci. **180**(980), 40 (1942)
158. A. Immirzi et al., Crystal structure and melting entropy of natural rubber. Macromolecules **38**(4), 1223–1231 (2005)
159. Y. Takahashi, T. Kumano, Crystal structure of natural rubber. Macromolecules **37**(13), 4860–4864 (2004)
160. S.C. Nyburg, A statistical structure for crystalline rubber. Acta Crystallogr. A **7**(5), 385–392 (1954)
161. G. Natta, P. Corradini, über die Kristallstrukturen des 1,4-cis-polybutadiens und des 1,4-cis-polyisoprens. Angew. Chem. **68**(19), 615–616 (1956)
162. W. Yau, R.S. Stein, Optical studies of the stress-induced crystallization of rubber. J. Polym. Sci. A-2 Polym. Phys. **6**(1), 1–30 (1968)
163. L.A. Wood, N. Bekkedahl, Crystallization of unvulcanized rubber at different temperatures. J. Appl. Phys. **17**, 362 (1946)
164. E.N. Dalal, K.D. Taylor, P.J. Phillips, The equilibrium melting temperature of cis-polyisoprene. Polymer **24**(12), 1623–1630 (1983)
165. B. Huneau, Strain-induced crystallization of natural rubber: a review of X-ray diffraction investigations. Rubber Chem. Technol. **84**(3), 425–452 (2011). doi:10.5254/1.3601131. http://link.aip.org/link/RCT/84/425/1
166. E.H. Andrews, Spherulite morphology in thin films of natural rubber. Proc. R. Soc. Lond. Ser. A Math. Phys. Sci. **270**(1341), 232–241 (1962)
167. E.H. Andrews, Crystalline morphology in thin films of natural rubber. II. Crystallization under strain. Proc. R. Soc. Lond. Ser. A Math. Phys. Sci **277**(1371), 562–570 (1964)
168. S. Kawahara et al., Crystal nucleation and growth of natural rubber purified by deproteinization and trans-esterification. Polym. J. **36**(5), 361–367 (2004)
169. D. Luch, G.S.Y. Yeh, Morphology of strain-induced crystallization of natural rubber. Part II. X-ray studies on cross-linked vulcanizates. J. Macromol. Sci. B **7**(1), 121–155 (1973)
170. L. Mandelkern, *Crystallization of Polymers Volume 1: Equilibrium Concepts* (Cambridge University Press, Cambridge, 2004)
171. E.H. Andrews, P.J. Owen, A. Singh, Microkinetics of lamellar crystallization in a long chain polymer. Proc. R. Soc. Lond. Ser. A Math. Phys. Sci. **324**(1556), 79–97 (1971)
172. G.S.Y. Yeh, Strain-induced crystallization II. Subsequent fibrillar-to-lamellar transformation. Polym. Eng. Sci. **16**(3), 145–151 (1976)
173. G.S.Y. Yeh, Morphology of strain-induced crystallization of polymers. Part I. Rubber Chem. Technol. **50**, 863–873 (1977)
174. T. Shimizu et al., TEM observation of natural rubber thin films crystallized under molecular orientation. Rubber Chem. Technol. **73**(5), 926–936 (2000)
175. J.H. Magill, Crystallization and morphology of rubber. Rubber Chem. Technol. **68**, 507–539 (1995)

176. W.F. Reichert, M.K. Hopfenmüller, D. Göritz, Volume change and gas transport at uniaxial deformation of filled natural rubber. J. Mater. Sci. **22**(10), 3470–3476 (1987)
177. S.R. Ryu, J.W. Sung, D.J. Lee, Strain-induced crystallization and mechanical properties of NBR composites with carbon nanotube and carbon black. Rubber Chem. Technol. **85**(2), 207–218 (2012)
178. S. Trabelsi, P.A. Albouy, J. Rault, Crystallization and melting processes in vulcanized stretched natural rubber. Macromolecules **36**(20), 7624–7639 (2003)
179. M.F. Acken, W.E. Singer, W. Davey, X-ray study of rubber structure. Ind. Eng. Chem. **24**(1), 54–57 (1932)
180. H. Hiratsuka et al., Measurement of orientation crystallization rates of linear polymers by means of dynamic X-ray diffraction technique. II. Frequency dispersion of strain-induced crystallization coefficient of natural rubber vulcanizates in subsonic range. J. Macromol. Sci. B **8**(1-2), 101–126 (1973). doi:10.1080/00222347308245796. eprint: http://www.tandfonline.com/doi/pdf/10.1080/00222347308245796, http://www.tandfonline.com/doi/abs/10.1080/00222347308245796
181. H. Kawai, Dynamic X-ray diffraction technique for measuring rheo-optical properties of crystalline polymeric materials. Rheol. Acta **14**(1), 27–47 (1975). ISSN: 0035-4511, http://dx.doi.org/10.1007/BF01527209
182. N. Candau et al., Characteristic time of strain induced crystallization of crosslinked natural rubber. Polymer **53**(13), 2540–2543 (2012)
183. I. Temizer, P. Wriggers, T.J.R. Hughes, Contact treatment in isogeometric analysis with NURBS. Comput. Methods Appl. Mech. Eng. **200**(9), 1100–1112 (2011)
184. D.J. Dunning, P.J. Pennells, Effect of strain on rate of crystallization of natural rubber. Rubber Chem. Technol. **40**, 1381 (1967)
185. J.C. Mitchell, D.J. Meier, Rapid stress-induced crystallization in natural rubber. J. Polym. Sci. A-2 Polym. Phys. **6**(10), 1689–1703 (1968)
186. M. Tosaka et al., Crystallization and stress relaxation in highly stretched samples of natural rubber and its synthetic analogue. Macromolecules **39**(15), 5100–5105 (2006)
187. M. Tosaka et al., Detection of fast and slow crystallization processes in instantaneously-strained samples of cis-1,4-polyisoprene. Polymer **53**(3), 864–872 (2012)
188. L. Mandelkern, *Crystallization of Polymers Volume 2: Kinetics and Mechanisms* (Cambridge University Press, Cambridge, 2004)
189. P.G. De Gennes, Coil-stretch transition of dilute flexible polymers under ultrahigh velocity gradients. J. Chem. Phys. **60**(12), 5030–5042 (1974)
190. R.H. Somani et al., Flow-induced shish-kebab precursor structures in entangled polymer melts. Polymer **46**(20), 8587–8623 (2005)
191. J.F. Sanders, J.D. Ferry, R.H. Valentine, Viscoelastic properties of 1,2- polybutadiene: comparison with natural rubber and other elastomers. J. Polym. Sci. A-2 Polym. Phys. **6**(5), 967–980 (1968)
192. M. Rubinstein, R.H. Colby, *Polymer Physics* (Oxford University Press, Oxford, 2003)
193. L. Mandelkern, Crystallization kinetics of homopolymers: overall crystallization: a review. Biophys. Chem. **112**(2–3), 109–116 (2004)
194. C. Fond, Cavitation criterion for rubber materials: a review of void-growth models. J. Polym. Sci. B Polym. Phys. **39**(17), 2081–2096 (2001)
195. A. Dorfmann, K.N.G. Fuller, R.W. Ogden, Shear, compressive and dilatational response of rubberlike solids subject to cavitation damage. Int. J. Solids Struct. **39**(7), 1845–1861 (2002)
196. A.N. Gent, P.B. Lindley, Tension flaws in bonded cylinders of soft rubber. Rubber Chem. Technol. **31**(2), 393–394 (1958)
197. P. Adriaensens et al., Relationships between microvoid heterogeneity and physical properties in cross-linked elastomers: an NMR imaging study. Macromolecules **33**(19), 7116–7121 (2000)
198. J.B. Le Cam, A review of volume changes in rubbers: the effect of stretching. Rubber Chem. Technol. **83**, 247 (2010)

199. A.N. Gent, Cavitation in rubber: a cautionary tale. Rubber Chem. Technol. **63**(3), 49–53 (1990)
200. R. Christensen, C. Hoeve, Theoretical and experimental values of the volume changes accompanying rubber extension. Rubber Chem. Technol. **43**, 1473–1481 (1970)
201. R.W. Penn, Volume changes accompanying the extension of rubber. J. Rheol. **14**, 509–517 (1970)
202. T. Shinomura, M. Takahashi, Volume change measurements of filled rubber vulcanizates under stretching. Rubber Chem. Technol. **43**, 1025–1035 (1970)
203. R. Shuttleworth, Volume change measurements in the study of rubber-filler interactions. Eur. Polym. J. **4**(1), 31–38 (1968)
204. S. Kaneko, J.E. Frederick, D. McIntyre, Void formation in a filled SBR rubber determined by small-angle X-ray scattering. J. Appl. Polym. Sci. **26**(12), 4175–4192 (1981)
205. J. Ramier et al., In situ SALS and volume variation measurements during deformation of treated silica filled SBR. J. Mater. Sci. **42**(19), 8130–8138 (2007)
206. K. Layouni, L. Laiarinandrasana, R. Piques, Compressibility induced by damage in carbon black reinforced natural rubber, in *Proceedings of the III. European Conference on Constitutive Models for Rubber*, ed. by J.C.C. Busfield, A.H. Muhr (Taylor & Francis, London, 2003), p. 273
207. J.B. Le Cam, E. Toussaint, Cyclic volume changes in rubber. Mech. Mater. **41**(7), 898–901 (2009)
208. J.B. Le Cam, E. Toussaint, Volume variation in stretched natural rubber: competition between cavitation and stress-induced crystallization. Macromolecules **41**(20), 7579–7583 (2008)
209. T. Shinomura, M. Takahashi, Morphological study on carbon black loaded rubber vulcanizate under stretching. Rubber Chem. Technol. **43**, 1015–1024 (1970)
210. E. Bayraktar et al., Damage mechanisms in natural (NR) and synthetic rubber (SBR): nucleation, growth and instability of the cavitation. Fatigue Fract. Eng. Mater. Struct. **31**(2), 184–196 (2008)
211. P.A. Kakavas, W.V. Chang, Acoustic emission in bonded elastomer discs subjected to compression. J. Appl. Polym. Sci. **45**(5), 865–869 (1992)
212. J.C. Goebel, A.V. Tobolsky, Volume change accompanying rubber extension. Macromolecules **4**(2), 208–209 (1971)
213. J.M. Chenal et al., Parameters governing strain induced crystallization in filled natural rubber. Polymer **48**(23), 6893–6901 (2007)
214. A.N. Gent, B. Park, Failure processes in elastomers at or near a rigid spherical inclusion. J. Mater. Sci. **19**(6), 1947–1956 (1984)
215. A.N. Gent, P.B. Lindley, Internal rupture of bonded rubber cylinders in tension. Proc. R. Soc. Lond. Ser. A Math. Phys. Sci. **249**(1257), 195–205 (1959)
216. K. Cho, A.N. Gent, Cavitation in model elastomeric composites. J. Mater. Sci. **23**(1), 141–144 (1988)
217. A.N. Gent, D.A. Tompkins, Nucleation and growth of gas bubbles in elastomers. J. Appl. Phys. **40**, 2520 (1969)
218. P. Kumar et al., Volume changes under strain resulting from the incorporation of rubber granulates into a rubber matrix. J. Polym. Sci. B Polym. Phys. **45**(23), 3169–3180 (2007)

Chapter 2
Motivation and Objectives

The study of structure-property relationships is at the heart of polymer science, especially polymer physics and engineering. Filler-reinforced natural rubber is a particularly interesting example, because it exhibits several phenomena, which are scientifically challenging as well as industrially important. Although strain-induced crystallization was discovered almost a century ago and has long been known to be responsible for the outstanding mechanical and tear fatigue properties of natural rubber, the details are still unclear. Unlike in thermal crystallization, the kinetic mechanism of strain-induced crystallization is still an open question. Numerous studies attempted to shed light on the strain and temperature dependence of strain crystallization, but the time dependence has seen little research, despite the fact that rubber products are often subjected to dynamic mechanical loads (e.g. tires). The shortcomings in the characterization are attributed to the experimental difficulties to study the material structure on short time scales. The preferred technique to study crystallization is wide-angle X-ray diffraction, typically a time-consuming method. Only since the construction of ever more powerful synchrotron radiation sources over the last 10–15 years, the insight into these areas has become experimentally feasible.

Besides the self-reinforcement by strain-induced crystallites, the majority of industrially relevant rubbers are reinforced with particulate fillers, like carbon black or silica. Numerous models exist to describe the reinforcing ability of carbon black, putting forward different reinforcing mechanisms. Despite the enormous industrial relevance, till now no model can predict the viscoelastic behavior of an elastomer composite by first principles.

The focus of this work is to design and perform experiments to gain new insight into the short-time regime of strain-induced crystallization and to establish new relationships between structure and mechanical behavior. Spatially resolved wide-angle X-ray diffraction serves to study the structural evolution around a crack tip, preferably under fatigue conditions, to obtain an in-depth understanding of tear fatigue, crack propagation and failure of rubber parts.

K. Brüning, *In-situ Structure Characterization of Elastomers during Deformation and Fracture*, Springer Theses, DOI: 10.1007/978-3-319-06907-4_2,

The role of the filler structure and orientation under mechanical load is investigated and set in relation with local deformation in the matrix to give new insight into the reinforcing mechanisms of filled rubbers. Cavitation, widely discussed as a factor influencing the reinforcing mechanism due to its dissipative properties, but also regarded as a potential source of material failure, is studied with novel optical methods as well as with ultra-small angle X-ray scattering.

Scattering experiments are complemented by scanning electron microscopy analysis. An in-situ tensile machine was designed to observe the growth of a crack in real time on the size scale of 100 nm.

Deformation and fracture phenomena in rubbers can be studied with various techniques, each accessing the issue from a different direction and thus contributing to the overall goal, which is to obtain a complete understanding of the deformation and fracture behavior, all the way from the molecular basis to the implications for the rubber and tire industry. To gain this understanding, it is desirable to obtain in-situ quantitative and representative information about the processes occurring in the bulk and in the vicinity of the crack tip with good spatial and time resolution. Since classical techniques like electron microscopy often are limited to surface information and also cannot provide real time information, the methods of choice for these investigations are X-ray scattering and diffraction.

Especially since the advent of powerful synchrotron sources over the last two decades, numerous researchers have taken advantage of small-angle X-ray scattering (SAXS) and wide-angle X-ray diffraction (WAXD) techniques to study structural changes in polymers under deformation. By the combination of SAXS and WAXD, structural changes of various length scales can be observed, ranging from a few Å to almost a μm. Thus processes like the orientation of the network under deformation, the ordering of chain segments into crystallites, the alignment of anisometric filler particles along the tensile direction, cavitation and void formation are directly accessible. The major drawback of scattering techniques is that all information is provided in reciprocal space, i.e. in order to be accessible to common human perception, either a transformation to real space has to be done or certain quantitative features have to be extracted directly from reciprocal space.

The goal of this work was to study the effect of strain field and time on the structure of the rubber. Natural rubber (NR) possesses the outstanding ability to undergo strain-induced crystallization (SIC). The crystallites are generated in highly stretched regions and are oriented along the direction of the principal strain. They considerably elevate the tensile strength of the rubber, acting as reinforcing rigid particles and additional crosslinking points at the same time. This feature sets natural rubber apart from synthetic rubbers and is the reason for its excellent tear resistance, since the crystallites in front of the crack tip impede the propagation of the crack through the reinforced zone.

In the context of SIC, WAXD was used to study the crystallization under uniaxial and biaxial load. Relations between crystallinity, strain, filler type and filler content were established. Using a micron-size X-ray beam, the local crystallinity around a crack tip was studied under static conditions as well as under dynamic conditions.

The kinetics of the crystallization process was studied with an unprecedented time resolution in the ms range. This was achieved by using a high-flux third generation synchrotron source.

To make the step to the filler structure, the degree of crystallinity was used as an indicator for the local strain in the matrix in filled rubbers and thus a strain amplification factor was derived. The orientation of fillers was followed by SAXS and USAXS (ultra small-angle X-ray scattering). Considering the complex structure of industrial fillers like carbon black and silica and the associated difficulties in the interpretation of the scattering data, model fillers of similar size, but with a spherical and rod-like geometry were used additionally. Advantage was taken of the crystalline nature of these fillers and a model was developed for a quantitative interpretation of the SAXS data.

Due to the formation of new phase boundaries, USAXS and SAXS also are ideally suited to quantitatively study the formation of cavities, which are hypothesized to serve as nuclei for a possible macroscopic failure. These results were set in relation to macroscopic volumetric studies, using a newly developed optical method. Scanning electron microscopy during in-situ tensile tests was used to complement the scattering data with real space information.

In the next chapter, the reader will be familiarized with the characteristics of natural and synthetic rubbers, the use of fillers for the reinforcement of rubbers, and with the demands that are put on rubber products, e.g. mechanical strength and tear resistance. The theoretical backgrounds behind the methodology to study these properties and to establish new structure-property relationships are laid out. After a brief introduction to the nature of X-rays, the fundamentals of scattering and diffraction techniques are introduced, especially in the context of synchrotron facilities. In Chap. 3, the focus is on the experimental part. The sample preparation and their mechanical characterization are given as well as a detailed description of the numerous experimental setups. In Chap. 4, the outcomes of various scattering and diffraction experiments under static and dynamic conditions are presented. In addition, it shows the results of volumetric experiments and electron microscopy studies. The thesis finishes with a conclusion and an outlook onto future prospects in the field.

This work was done in the framework of the DFG-Forschergruppe FOR597 *Fracture Mechanics and Statistical Mechanics of Reinforced Elastomeric Blends*, consisting of seven projects carried out at five partner institutes throughout Germany, with a funding period of six years. The project *Online Structure Characterization of Elastomer Compounds During Deformation and Fracture* was established after completion of the first three years of the research group.

Chapter 3
Experimental

3.1 Materials and Samples

In order to investigate the influence of different parameters on the strain-induced crystallization and other structural changes, the following quantities were varied: matrix type, crosslinking density, crosslinking type, filler content and filler type. Various rubber recipes were compounded, vulcanized and subjected to mechanical tests and structural characterizations. While initially the green compounds were provided by the Deutsches Institut für Kautschuktechnologie (DIK, Hannover) in the framework of the FOR597 research group, the majority of the samples were prepared in the elastomers laboratory at IPF.

A detailed list of all recipes, that are discussed in this work, is to be found in Sect. A.6. When dealing with a certain material, the main characteristics are given in the text, and the reader is referred via the material ID [#...] to the list for details.

3.1.1 Compounding

Compounding was carried out in two steps. First, all ingredients (except the crosslinker and accelerator) were mixed in an internal mixer (Haake Rheocord PolyLab 300p with HAAKE Rheomix 600p, chamber volume 78 mL, typically 7 min at 70 rpm and 70 °C) (Fig. 3.1a). Afterwards, the crosslinker was added on a two-roll mill (Servitec Polymix 110 L, 45 °C roll set temperature, 0.5 mm gap size, roll speeds 14 and 9 mm/s, 15 passes) (Fig. 3.1b). Mixing on the two-roll mill is preferred over the internal mixer because of the better homogenization and higher shear forces, which especially aids in the dispersion of low crosslinker contents.

Usually, to avoid aging effects as well as the presence of impurities like antioxidants, new series of materials were mixed and vulcanized roughly two weeks before each beamtime, and then stored refrigerated at −20 °C until a few days before being tested.

K. Brüning, *In-situ Structure Characterization of Elastomers during Deformation and Fracture*, Springer Theses, DOI: 10.1007/978-3-319-06907-4_3,

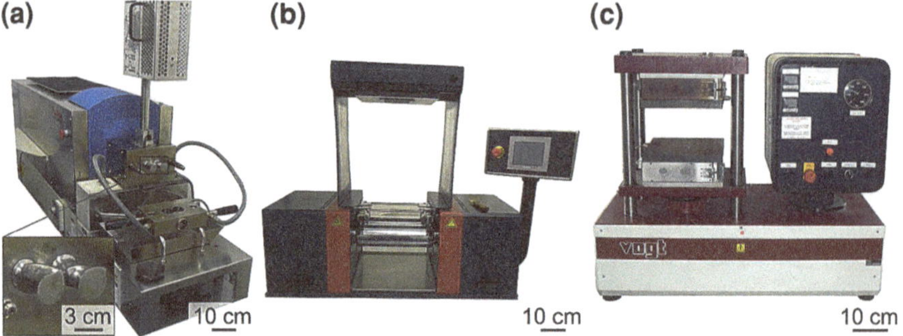

Fig. 3.1 a Internal mixer; b two-roll mill; c hot press

3.1.2 Vulcanization

In order to optimize the vulcanization conditions, vulcameter tests were performed on a Scarabeus SIS-V50. The curing temperature was usually set to 160 °C. The vulcanization time was taken as t_{90} and additional time was added depending on the sample thickness and mold geometry. Vulcanization was carried out in a hot press (Fig. 3.1c). The mold was preheated to the vulcanization temperature before being loaded. A combined mold was used for pure shear specimens and strips (Fig. 3.2a); and separate molds were used for cruciform-shaped biaxial samples and for dumb-bell samples for optical volumetric studies (Fig. 3.2b). The dumb-bell mold has a vertical parting plane. It was designed to be used in a heated tablet press, pressing along the symmetry axis, to maximize the hydrostatic pressure in the mold, reduce dead zones and thus avoid the presence of production-related voids. Furthermore, simple cylindrical specimens were obtained from the tablet press.

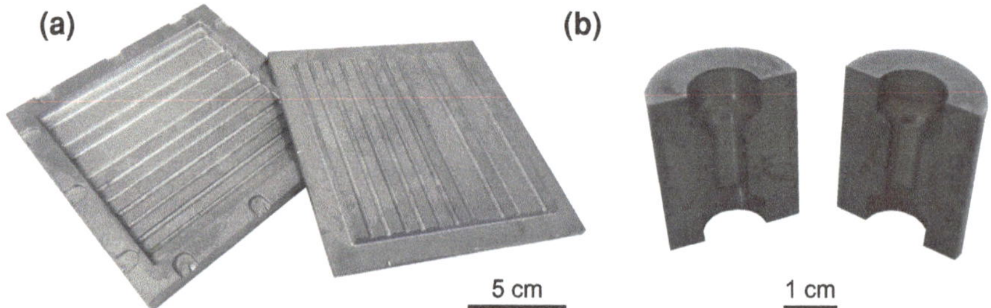

Fig. 3.2 Vulcanization molds for a strips to be used for die-punching uniaxial tensile bars and cutting pure shear samples; b dumbbell specimens

3.1.3 Sample Geometry

Tensile samples were obtained by die-punching of the vulcanized rubber piece. The geometry of the tensile bars is shown in Fig. 3.3a. The beads in the clamping area minimize sample slippage in the grips. Biaxial tests were performed on a cruciform-shaped specimen, which is shown as obtained from the mold without further cutting or die-punching in Fig. 3.3b. The pure shear specimen for pure shear tensile tests and tear fatigue experiments were obtained by cutting of 100 mm wide strips (Fig. 3.3c). The dumb-bell samples are rotation symmetric as shown in (Fig. 3.3d).

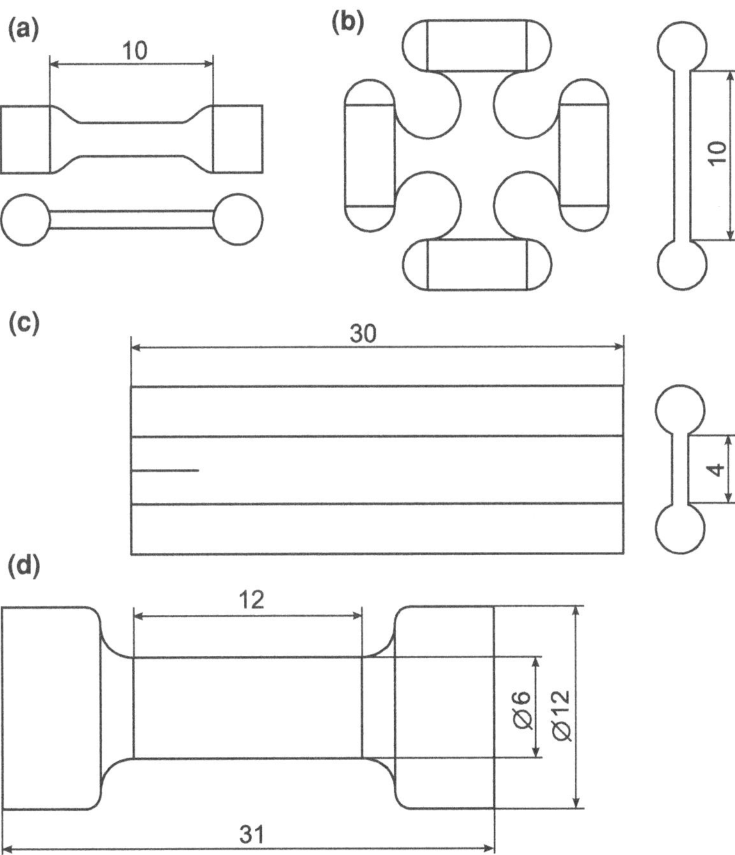

Fig. 3.3 **a** Uniaxial tensile bar; **b** biaxial specimen; **c** notched pure shear specimen for tear fatigue experiments; **d** dumbbell specimen for volumetric studies

3.2 Synchrotron Experiments

Almost all of the scattering and diffraction experiments described in this work were carried out at the DESY (Deutsches Elektronen Synchrotron).[1] Beamtime was initially available in the framework of different long-term proposals. Later, own proposals were submitted and beamtime was granted. While the first half of the work was done at the wiggler beamline BW4 at the DORIS II ring, the construction and commissioning of the PETRA III third generation synchrotron light source in 2011 opened new possibilities. Friendly user beamtime was granted in spring 2011 before the official opening for regular user operation. All PETRA III experiments were done at the MiNaXS beamline.

3.2.1 Beamline Setup

The principle beamline setup at the BW4 and MiNaXS beamlines is depicted in Fig. 3.4. A photo is shown in Fig. 3.5. The adjustment of the X-ray energy and beam geometry is usually done by the beamline staff and thus is covered in the theory part (Sect. 1.2.1). After the beam exits the final aperture, its intensity is measured by an ionization chamber.

The ionization chamber is especially critical at the BW4 beamline, because the beam intensity considerably decreases over the period of one run (six hours), before positron injection into the storage ring re-establishes the previous beam intensity. At PETRA III, injection is done continuously and beam intensity can be taken to be constant. Thus, no ionization chamber is used.

The sample is placed as close as possible to the final aperture or the ionization chamber, respectively, in order to minimize air scattering. The sample size should

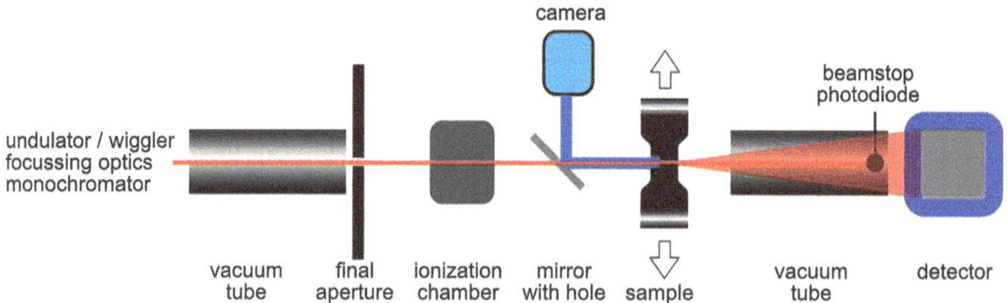

Fig. 3.4 Sketch of typical beamline setup. For WAXD, no vacuum tube is used between sample and detector. At MiNaXS, the ionization chamber is not available. For time-resolved experiments, no photodiode was used. The camera and the mirror are only applied in experiments with the quasistatic tensile machine

[1] Few preparatory experiments were done at a lab source. This will not be described in detail.

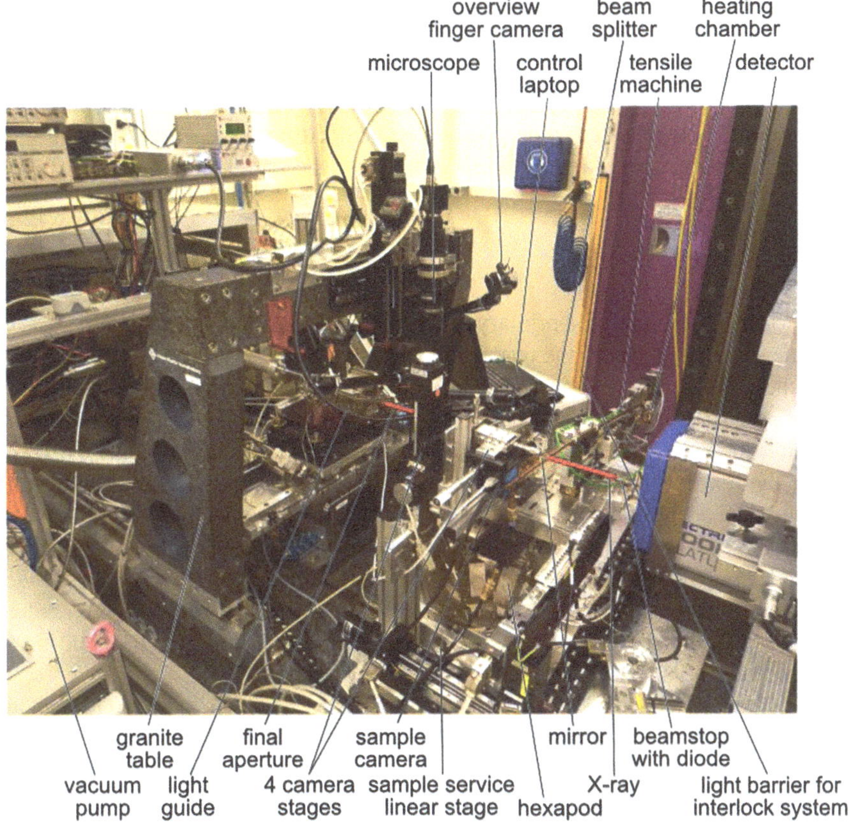

Fig. 3.5 Photograph of the quasistatic tensile experiments setup in the MiNaXS beamline. The tensile machine is currently moved out of the beam to change the sample. The beam is drawn in *red*

be chosen such that its thickness gives a good balance between beam absorption and number of scatterers. Simple geometrical considerations yield for the optimum sample thickness t_{opt} [1]:

$$t_{opt} = \frac{\cos(2\Theta)}{\mu}, \tag{3.1}$$

where μ is the absorption coefficient, which varies with wavelength. For unfilled and filled rubbers, the optimum sample thickness is in the range between 2 and 7 mm.

Since the sample is mounted in a tensile machine and the precise sample alignment is critical, the mechanism to change the sample has to be accurate and easy to handle at the same time, in order to reduce deadtime. At BW4, an in-house made manual sliding carriage with an adjustable bedstop is used. At MiNaXS, the beamline provides an electric stage, moving the complete sample environment perpendicularly out of the beam.

Behind the sample, a beamstop blocks the primary beam, as it would otherwise damage the detector. Typically, the beamstop is made from lead. Unless an extremely small beamstop is required, a photodiode is included in the beamstop. Along with the ionization chamber readouts, one can calculate the sample transmission τ:

$$\tau = \frac{I_{s,2}}{I_{s,1}} \frac{I_{bg,1}}{I_{bg,2}}. \tag{3.2}$$

Here, I is the intensity of the primary beam, measured in front of the sample ($I_{s,1}$) or after the sample ($I_{s,2}$), and the respective values without sample are $I_{bg,1}$ and $I_{bg,2}$.

At the end of the setup, one finds the detector. Nowadays, almost exclusively two-dimensional CCD detectors are used. The X-ray photons are taken up by a fluorescent film, emitting light in the visible range, which is finally recorded by a CCD-chip. The small-angle experiments at BW4 employed a MAR 165 detector. Here, the light from the individual array elements of the fluorescent screen is guided by optical fibers to a smaller CCD-chip, which is then read out by a computer. On the contrary, in the more recent CCD-detectors, commonly called *pixel detectors*, a small CCD-chip resides behind each pixel of the screen. Consequently, pixel detectors provide zero readout noise, a higher dynamic range and shorter readout times. PSI Pilatus 100, 300 K and 1 M pixel detectors were used.

To avoid air scattering, a vacuum tube is placed between the sample and the detector in the SAXS and USAXS setups.

The hitherto described part of the beamline is placed in the experimental hutch. In order to shield the users from the X-ray radiation, the hutch may not be entered while the X-ray shutter is open. The complete experimental control—including the operation of the tensile machines—has to be done remotely and via cameras. To achieve this, the beamline is equipped with a control stand featuring numerous computers. To allow for a flexible setup (e.g. change between WAXD, SAXS and USAXS, or accomodating different sample environments), almost all stages are electrically driven. For example, one can vary the flight tube length, or adjust the sample position in any spatial direction. The computers also control the detectors. The tensile machines are remote controlled as well. Only for changing the samples, the user is required to break the interlock and enter the hutch.

Before the scattering data can be analyzed, masking of irrelevant pixels, background correction and normalization need to be performed. Masking is necessary to exclude non-physical pixel values (e.g. caused by the beamstop, the beamstop holder, or blind detector areas). Optionally, the masked areas can be restored from symmetry conditions (point symmetry or mirror symmetry) after the background correction (Fig. 3.6). For background correction, patterns without sample, I_{bg}, are acquired. The background scattering is subtracted from the sample pattern, I_s, to obtain the background-corrected pattern, I_c:

$$I_c = I_s - \tau I_{bg} \tag{3.3}$$

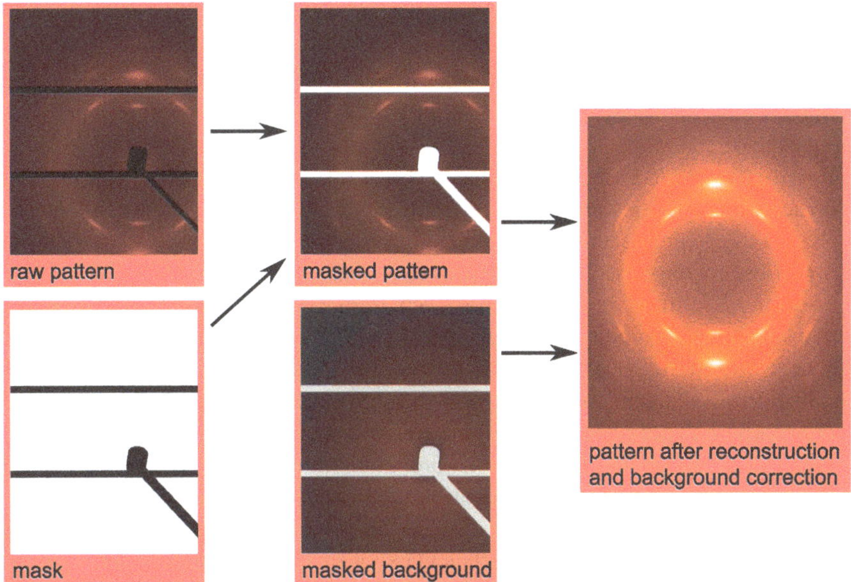

Fig. 3.6 Masking, background correction and reconstruction of a WAXD pattern, exemplified on filled NR [#10] uniaxially stretched to 450 %. Stretching direction is horizontal. The blind areas (*black bars*) on the detector are due to the segmental architecture of the Pilatus 300 K design. The color saturation of the background was enhanced for illustration reasons

Since the thickness of the sample varies with strain, the pattern is then normalized with respect to unit sample volume:

$$I_n = I_c \frac{-1}{\tau \ln(\tau)} \tag{3.4}$$

In the following, the different beamline setups will be described individually, followed by a section dealing with the data postprocessing.

3.2.2 Quasistatic Tensile Experiments

In-situ quasistatic tensile experiments were carried out on an in-house (IPF Forschungstechnik) made miniature tensile testing device, specifically designed for the application in beamlines:

- The beam can penetrate the sample.
- The machine does not interfere with the diffracted radiation.
- The maximum displacement is sufficient to study elastomers up to fracture—a common shortcoming of commercially available miniature tensile machines. The

machine allows a maximum displacement of 100 mm and a maximum force of 250 N.

- The grips move symmetrically in order to keep the center of the tensile sample aligned with the beam. This is ensured by two linear electric motors running in opposite directions.
- The machine dimension along the beam is minimized to reduce air scattering. The vertical dimension is compact to allow fitting of electric stages between the base plate of the beamline and the machine.
- A pull-out carriage mechanism ensures easy change of samples.
- Rotation and tilting around two axes, each perpendicular to the beam, allows to check the fiber symmetry and to explore the reciprocal space.
- In order to obtain optical strain data, a camera observes the sample along the beam via a mirror with a hole for the primary beam, placed at 45° to the beam.
- The machine is fully operable by remote control from the control room.
- Coupling of the tensile machine software with the beamline and detector control computers allows automatic acquisition of a pattern series in the stretch-hold-technique, i.e. patterns are acquired in certain strain intervals, during which the machine is stopped to allow for sufficient exposure time. The names of the photos and patterns are synchronized and stored in a table along with displacement, force and time.

The machine is depicted in Fig. 3.7. When used at MiNaXS, the machine was operated in continuous mode rather than in the stretch-hold technique owing to the shorter exposure times.

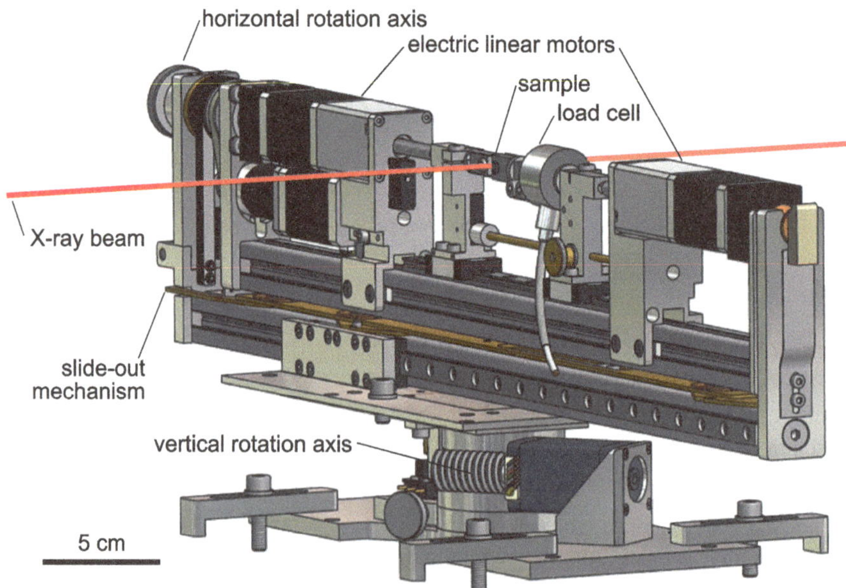

Fig. 3.7 Miniature tensile testing machine for quasistatic tensile tests at the synchrotron

3.2.3 Cyclic Dynamic Tensile Experiments

Since the maximum deformation rate of the quasistatic tensile machine is limited to 2 mm/s, a dynamic tensile machine was designed to study structural evolution under mechanical conditions equivalent to tear fatigue experiments, i.e. at frequencies around 1 Hz. An electric motor drives an eccentric and a lever mechanism translates the motion into a symmetric stretching of the tensile sample (Fig. 3.8). The stroke can be adjusted as well as the pre-strain. The machine was intended to serve as a prototype, but turned out to work well and delivered good results.

An optical strain versus time record, obtained by analysis of single frames of a video film (30 fps), is shown in Fig. 3.9. Despite the simple design without a step motor, the strain curve well resembles a sinusoidal curve. The frequency is 1.1 Hz. During the in-situ experiments, the strain was not permanently recorded. Only the minimum and maximum strains were recorded before turning on the machine, and the total scattering intensity of the WAXD patterns was used to assign a strain value to each pattern. The experiments were performed at MiNaXS. WAXD patterns were recorded over a period of 100 s with a frame rate of 50 fps (17 ms exposure time and 3 ms dead time) on a Pilatus 300 K detector. Several slit systems were fully opened to improve the photon rate by a factor of ten. This increases the beam size to roughly 100 μm, which is still smaller than at most other beamlines and is sufficiently small to avoid excessive peak broadening.

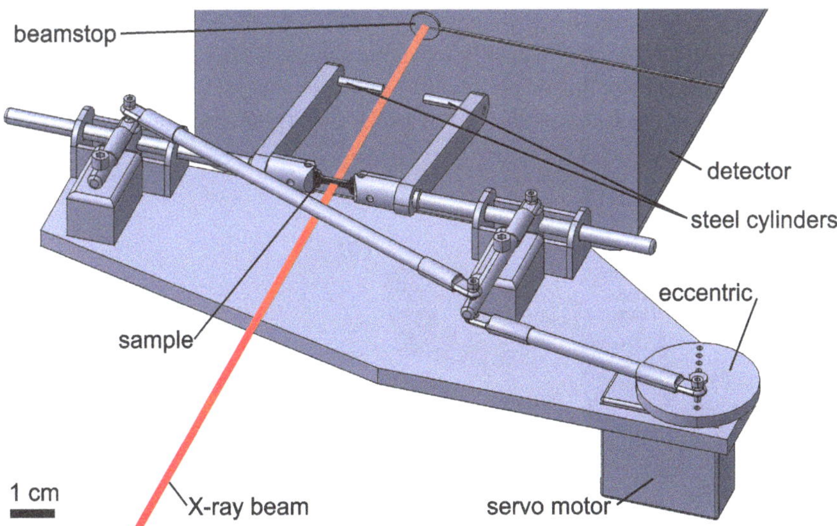

Fig. 3.8 Drawing of the dynamic tensile machine set up in the beamline. The aluminum levers with the steel cylinders are only used in the dynamic scanning experiments

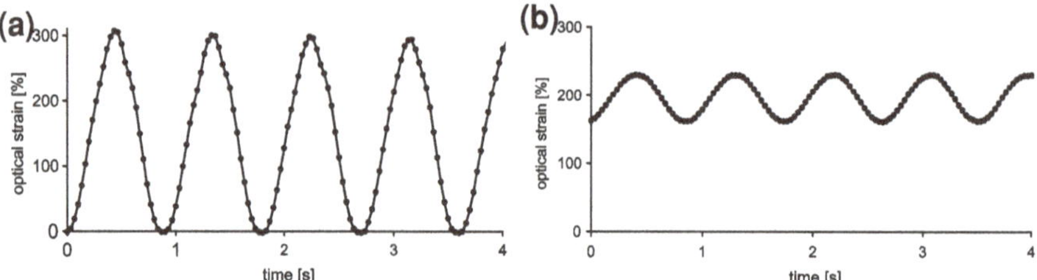

Fig. 3.9 Optical strain histories of stretching experiments with the dynamic tensile machine. **a** Fully relaxing with maximum strain of 300 %; **b** non-relaxing with minimum strain of 160 % and maximum strain of 230 %

3.2.4 Tensile Impact Experiments

In order to directly access the kinetics of the strain-induced crystallization, the intention was to perform strain-jump experiments in analogy to temperature-jump or pressure-jump experiments in chemical reaction kinetics analysis. Thus, the strain rate should be as high as possible. A spring-driven tensile machine (Fig. 3.10) was designed to stretch the rubber sample to a pre-defined strain within less than 10 ms, corresponding to a strain rate of more than $350\,s^{-1}$. The machine is not equipped with a load cell or displacement transducer, but it can be connected to the Instron Electropuls E 1000 load cell (with a time resolution of 0.5 ms), which served to verify the strain rate. The spring is released by a remote controlled servo motor. Tensioning of the spring has to be done manually when the sample is changed.

For the steplike unloading experiments, an adapter was used to mount the sample on the other side, such that with the release of the spring the strain was removed (Fig. 3.10b).

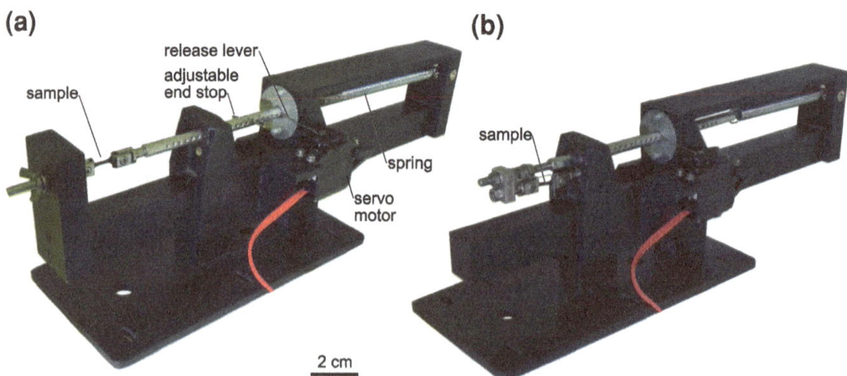

Fig. 3.10 Tensile impact machine for high strain rate tensile tests. The servo motor releases a spring that acts on the sample. **a** steplike loading setup; **b** steplike unloading setup

Experiments were performed at MiNaXS with a frame rate of 143 fps (4 ms exposure time and 3 ms dead time) on a Pilatus 300 K detector. Synchronization between the strain step and the WAXD patterns was ensured by a shadow of the detector, which moved when the spring was released.

3.2.5 Biaxial Tensile Experiments

In general, biaxial tensile experiments are carried out either on a square-shaped sample with segmented clamps along all four edges, or on cruciform samples with a conventional clamp on each of the for arms. Owing to the need to miniaturize the sample, only the cruciform geometry is feasible. A biaxial tensile machine was built by combining the quasistatic machine (Sect. 3.2.2) with an existing modified commercial miniature tensile machine (Kammrath and Weiss). The former operates the horizontal axis, while the latter takes care of the vertical axis. The two axes are controlled independently from different computers, but due to the very low strain rate (around 10 mm/min) of the vertical axis, synchronization is not an issue.

3.2.6 Static Crack Tip Scans

Static crack tip scans were performed on single edge notched tensile (SENT) samples.[2] The geometry of the latter is consistent with the classic tensile samples (Fig. 3.3a), with a notch of 0.5 mm depth introduced along the thickness direction in the middle of the sample. The sample is stretched following the stretch-hold technique (in steps of 1 mm), and during some of the hold intervals, a two-dimensional WAXD scan is performed. Other holding intervals served only to take a photo of the sample. Despite the beam size of roughly $35 \times 25\,\mu m$ at MiNaXS, a scan step width of $100\,\mu m$ in both directions was chosen to account for the finite thickness of the sample and for the supposedly imperfect alignment of the crack tip with the beam.[3] The exposure time was 0.5 s. The individual 2D scans were performed fully automatized in the Online interface provided by the beamline control computers. The vertical displacement was realized by a hexapod platform, while the sample service linear stage performed the horizontal movements. The 2D scan is composed of line scans in horizontal direction. To correlate the local structure obtained by the scans with the local strain field, a speckle pattern was applied to the sample to be evaluated ex-situ by Aramis digital image correlation software. Due to the necessity to record high-quality photographs without artefacts, the mirror setup could not

[2] For comparison, a small number of notched pure shear specimens (Fig. 3.3c) were scanned as well.

[3] In other words, one cannot be sure whether the root of the notch is perfectly aligned along the beam on a size scale of $20\,\mu m$.

be used. Instead, a second camera was placed parallel to the beam, around 20 cm displaced sideways. The sample was stretched while being in front of the second camera, and the moved into the beam by the sample service stage to start a scan.

3.2.7 Dynamic Crack Tip Scans

Dynamic crack tip scans were performed to obtain insights into the crystallinity evolution around the crack tip under conditions perfectly resembling tear fatigue experiments [2]. Due to the limited force range of the dynamic tensile machine, SENT samples were tested instead of pure shear samples. A further advantage of SENT samples is, that they can withstand larger strain amplitudes without crack propagation, such that a larger crystalline zone is to be expected. Preparatory experiments were done on the Instron machine to find the maximum amplitude, at which no crack growth takes place. A moderately carbon black-filled natural rubber was chosen because it gives sufficient resistance to crack growth during the extended cycling in the beamline and also the resistance to beam damage is superior to unfilled samples, which would not be able to withstand such a long accumulated exposure. A moderate filler content also allows to detects SIC at lower strains compared to the unfilled material, whilst reduction of the failure strain as in highly filled materials is avoided. A low crosslinking density was used to obtain high extensions and low stresses, considering the limited force range of the machine and the long test times.

The experimental procedure is a combination of the dynamic tensile tests and the static crack tip scans. Beam size and scan step size are the same as in the static experiment (Sect. 3.2.6). During dynamic scans, the dynamic tensile machine was operated continuously during a complete scan. This is in contrast to the setup recently proposed by Rublon et al. [3]; however a continuous operation is crucial in order to capture the effects of the finite crystallization kinetics. At each scanning point the stages halt for roughly 2 s, and a series of 100 patterns with a frame rate of 50 fps is acquired, with an exposure time of 17 ms per pattern. During these 2 s holding time, the tensile machine performs 2 strain cycles. The pattern acquisition and the tensile machine are not in phase. For one dynamic scan, approximately 50,000 patterns were acquired, corresponding to a total file size of 50 GB.

In the postprocessing data evaluation, each diffraction pattern is assigned a position (in the laboratory reference system) and a phase angle of the mechanic cycle. The position is easily obtainable from the scan devices' data output. The phase angle of the stretching cycle is obtained from the positions of shadows on the detector, cast by purposefully designed steel cylinders which are mounted to the grips of the tensile machine (Fig. 3.8). The distance between the shadows of the two steel cylinders is computed with sub-pixel precision iteratively with PV-Wave, using a Sobel algorithm for horizontal and vertical edge enhancement (Fig. 3.11). The distance between the shadows is then converted to the phase angle of the tensile machine. The maximum and minimum displacement can be read directly from the tensile machine. The gradient of distance over time reveals whether a pattern is to be assigned to

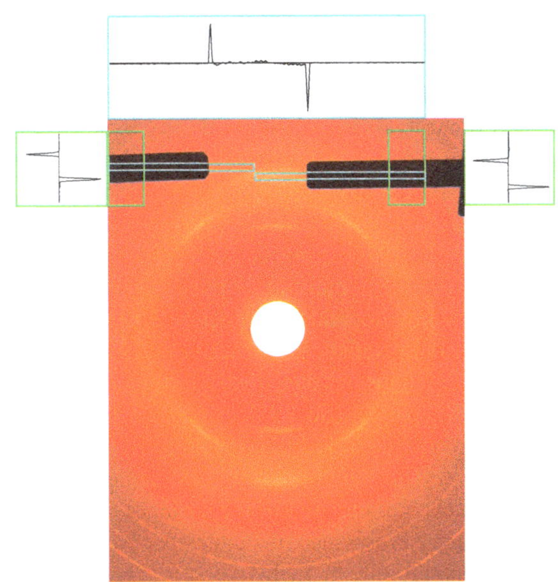

Fig. 3.11 Diffraction pattern including the shadows cast by the steel cylinders, allowing for a precise assignment of strain phase to each pattern. The *curves* represent the Sobel functions of the intensity in the marked areas (*green* for vertical position, *blue* for horizontal position)

the loading or unloading part of a cycle. To obtain a two-dimensional crystallinity map at a given phase angle, the strain cycle was discretized into ten intervals and each pattern was assigned to one of these intervals. A sequence of 10 crystallinity maps gives a video, showing the crystallinity evolution during fatigue cycling in real time.

The methodology requires the crack contour and the material structure to be constant over the number of cycles. The reproducibility of SIC under cyclic load was shown in previous work [4]. The constancy of the crack contour at maximum displacement was confirmed by two-dimensional static scans before and after the dynamic scans.

The motion of the sample during exposure introduces a spatial inaccuracy, rendering spatial resolutions better than 100 μm impossible despite the smaller beam. For instance, with an amplitude of 3.5 mm (fully relaxing with maximum displacement of 7 mm) the clamps move at a maximum rate of 370 μm/frame, whereas in the area of interest close to the crack tip the motion is only 100 μm/frame (at outer boundaries of the scanned area). Closer to the crack tip and at the extrema of a strain cycle, the inaccuracy introduced by the moving sample is negligible.

3.2.8 Crystallinity Determination by WAXD

In WAXD experiments, the scattering angle of interest is $5° < 2\Theta < 25°$, or in terms of scattering vectors, $6\,\text{nm}^{-1} < q < 27\,\text{nm}^{-1}$ (Sect. 1.2.4). The large angle implies a short detector distance, typically between 100 and 200 mm. For simple tensile experiments, advantage can be taken of several symmetry conditions. First, if the thickness and the height of the sample are similar, it fulfills fiber symmetry, i.e. it is rotation symmetric around the tensile axis. Thus, the scattering pattern is invariant to sample rotation about this axis and one pattern, representing a slice through reciprocal space, is sufficient to extract a relative measure for the degree of crystallinity. Second, if the tensile axis is placed horizontal and perpendicular to the beam and the detector is placed perpendicular to the beam, the pattern is symmetric to vertical and horizontal mirror planes. In other words, one quadrant contains the complete information, and the other quadrants can be seen as duplicates, improving the photon count statistics and thus enhancing the signal-to-noise ratio.

Various methods are described in literature to extract the degree of crystallinity from a WAXD pattern (Sect. 1.2.4). Because the time-resolved measurements require an accurate computation of crystallinity from relatively noisy data, the integration area of the crystallinity determination method should mainly include the most significant pixels, i.e. the area of the strongest diffraction peak, the (120) peak [5]. If fiber symmetry is assumed, the intensity of a single peak is sufficient to quantify the degree of crystallinity Φ. In this work, the crystallinity was defined to be proportional to the ratio of the integrated intensity of the (120) peak to the integrated intensity of the amorphous halo:

$$\Phi = k_1 \frac{\int_{14.3}^{15.6} \int_{-1.5}^{1.5} P_{cr} \, \mathrm{d}q_3 \mathrm{d}q_{12}}{\int_{8.7}^{19.4} \int_{-1.5}^{1.5} k_2 P_{am} \, \mathrm{d}q_3 \mathrm{d}q_{12}} = k_1 \frac{A_{cr}}{A_{am}}. \tag{3.5}$$

The integration limits in q-space[4] are given in nm^{-1} and were empirically defined to cover the corresponding ranges in the diffraction curve (Fig. 3.12c). Performing an integration along q_3 from -1.5 to $1.5\,\mathrm{nm}^{-1}$ corresponds to a slice in equatorial direction of roughly 50 px thickness, corresponding to an azimuthal integration angle of roughly 14°. The result was found to be insensitive to a variation of the slice thickness. The crystalline pattern P_{cr} was obtained by subtraction of the normalized amorphous halo $k_2 P_{am}$ from the actually measured intensity of the semicrystalline sample. The normalization scalar k_2 was adjusted such that the crystalline pattern P_{cr} does not show any intensity in regions unaffected by crystallization ($11.7\,\mathrm{nm}^{-1} < q_1 < 12.6\,\mathrm{nm}^{-1} \wedge -1.5\,\mathrm{nm}^{-1} < q_3 < 1.5\,\mathrm{nm}^{-1}$). The ratio was multiplied by an empiric constant k_1, in order to bring the relative ratio in line with absolute quantities found in literature, e.g. from DSC or density measurements. The computation steps are illustrated in Fig. 3.12. The described method was used for all kinds of WAXD setups and was shown to yield reproducible values when comparing reference sample results from different beamtimes with different detector setups.

[4] The q-space is commonly defined with q_3 being the tensile axis, and q_1 and q_2 being orthogonal to q_3. In a fiber-symmetric sample, $q_1 = q_2 = q_{12}$.

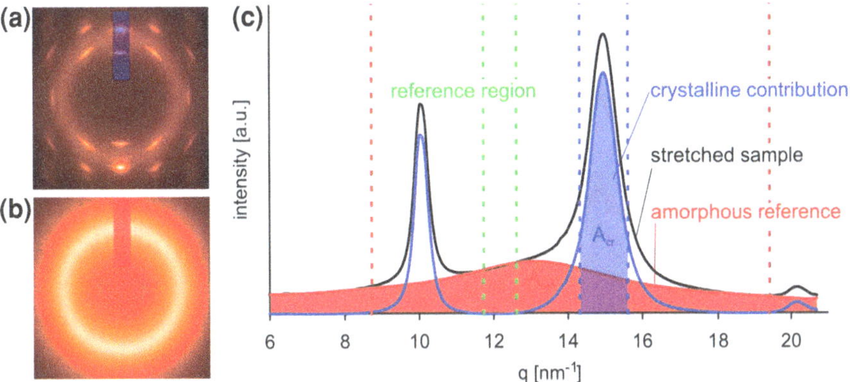

Fig. 3.12 Method to determine the degree of crystallinity Φ, exemplified on unfilled NR [#1] at 830 % strain. Φ = 14.5 %. **a** Semicrystalline pattern; **b** amorphous reference pattern, P_{am}. Both patterns are after background correction. The equatorial slices are extracted from the colored regions. **c** Equatorial slices of stretched sample (*black*), normalized reference (*red*) and crystalline contribution (*blue*), obtained by subtraction

3.2.9 Study of Filler Orientation by SAXS

In order to facilitate the understanding of the behavior of carbon black in a deformed matrix, it is advisable to start with model fillers to circumvent the problems associated with the evaluation of anisotropic carbon black scattering patterns (Sect. 1.1.3). The advantage of scattering experiments on geometrically simple scattering entities is that their scattering patterns can be calculated analytically. Further criteria for the selection of model fillers are their size, which should be similar to the size of carbon black aggregates, such that the structural changes can be observed at similar scattering angles and that the deformation of the matrix occurs on a comparable length scale. Moreover, a good scattering contrast between filler and matrix is desirable.

Given these requirements, nanosized aluminum oxide Al_2O_3 (Nanophase Nanodur) was chosen to represent spherical particles and aragonite (Schaefer Kalk), a form of precipitated calcium carbonate, was chosen to represent high aspect ratio cylindrical particles [6]. The beauty of these fillers is that they are monocrystalline substances, meaning their orientation can independently be followed by WAXD. Scanning electron micrographs of the raw fillers and after incorporation into the natural rubber matrix are shown in Fig. 3.13. The filler particles are sufficiently well dispersed and only few agglomerates are visible. The samples were crosslinked with dicumyl peroxide to avoid scattering from zinc oxide, which contributes to the SAXS and USAXS signals since its size is in the same range as that of the fillers [7].

In order to interpret the SAXS patterns, the information obtained from WAXD is particularly instructive. Under deformation, the aragonite diffraction rings turn into arcs and the orientational distribution of the diffraction intensity along the angle α can be well fit by a Maier-Saupe function [8–10]:

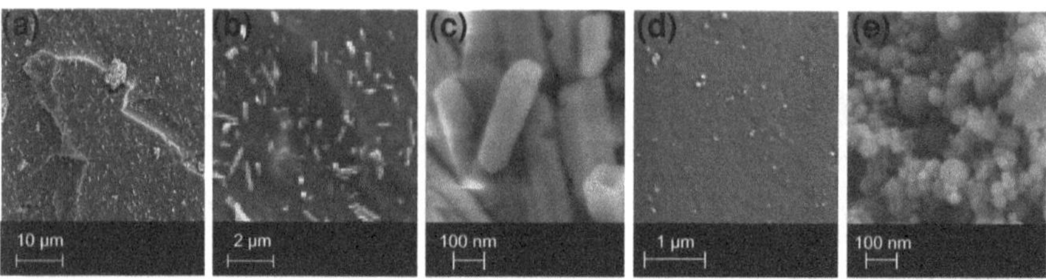

Fig. 3.13 Scanning electron micrographs of **a** and **b** NR with 27 phr aragonite (*scale bars* 10 and 2 μm), **c** aragonite powder (*scale bar* 100 nm), **d** NR with 10 phr Al_2O_3 (*scale bar* 1 μm), **e** Al_2O_3 powder (*scale bar* 100 nm) [6]

$$f_{MS}(\alpha) = \frac{1}{z}\exp\left(m\cos^2(\alpha)\right), \tag{3.6}$$

where $\frac{1}{z}$ is a normalization factor and m is the Maier-Saupe orientation parameter. With this information one can compute the USAXS patterns, if the geometry of the filler particles is known. Assuming a Gaussian size distribution of the cylinder length l, one obtains for the number density distribution $D(l)$:

$$D(l) = \frac{1}{\sigma_l\sqrt{2\pi}}\exp\left(\frac{-\left(l-\bar{l}\right)^2}{2\sigma_l{}^2}\right) \tag{3.7}$$

with the standard deviation σ_l and the average length $\bar{l}$. From SEM images, the length was taken as $\bar{l} = 600$ nm with a standard deviation $\sigma_l = 100$ nm. The aspect ratio a was kept fixed at 6. The size distribution was cut off at $\bar{l} \pm 3\sigma_l$.

The form factor of a cylinder, f_{cyl}, as a function of azimuthal angle α and scattering vector magnitude q is given by

$$f_{cyl} = k\frac{a\sin\left(q\frac{l}{2}\cos(\alpha)\right)J_1\left(q\frac{l}{2a}\sin(\alpha)\right)}{q^2l^2\cos(\alpha)\sin(\alpha)}, \tag{3.8}$$

where J_1 is the first order Bessel function of the first kind and k is a constant [11, 12]. Considering the moderate volume fraction of the aragonite filler (27 phr corresponds to 7.5 vol%), it is justified to approximate the structure factor as unity and take the scattering function of the polydisperse oriented ensemble simply as the sum of all individual contributions:

$$I_{cyl} = \int_{\alpha=0}^{\pi}\left(\left(1-\phi\right)f_{MS}(\alpha)+\phi\right)f_{cyl}(l,\alpha)^2 d\alpha, \tag{3.9}$$

where ϕ is the isotropic, unoriented portion of the particles, e.g. due to agglomeration. ϕ was taken as a fitting parameter and kept fixed at 0.2.

3.2.10 Cavitation Analysis by USAXS

Cavitation and void formation can be studied by a multitude of methods, as listed in Sect. 1.4. But only USAXS offers a bulk sensitivity with a good spatial resolution. In analogy to the work by Zhang et al. [13], the void fraction ϕ_{void} was derived from the change in total scattering intensity Q with respect to the initial total scattering intensity Q_0 of the undeformed sample. The total scattering intensity Q is defined as:

$$Q = \int_{0.03}^{1.0} \int_{0.03}^{1.0} I(q_1, q_3) q_3 \mathrm{d}q_1 \mathrm{d}q_3. \tag{3.10}$$

The integration limits are given in units of nm^{-1}. Theoretically, the integration extends over the complete reciprocal space. The limits given in Eq. 3.10 are set by experimental considerations.[5] Fiber symmetry is assumed. Theoretically, the integral extends from 0 to ∞, but clearly this is not experimentally feasible. No extrapolation of scattering data was attempted. The intensity $I(q_1, q_3)$ in Eq. 3.10 is after background correction and normalization to unit sample volume as described in Eqs. 3.3 and 3.4.

For a three-phase system (matrix, filler, voids), the relation between total scattering intensity and void fraction reads [14]:

$$\frac{Q}{Q_0} = 1 + \left(\frac{\phi_{\mathrm{r}} {\rho_{\mathrm{r}}}^2 + \phi_{\mathrm{f}} {\rho_{\mathrm{f}}}^2}{\phi_{\mathrm{r}} \phi_{\mathrm{f}} \left(\rho_{\mathrm{r}} - \rho_{\mathrm{f}}\right)^2} - 1 \right) \phi_{\mathrm{void}}. \tag{3.11}$$

Here, ρ_{r} is the scattering length density of the rubber phase ($8.756 \times 10^{10}\,\mathrm{cm}^{-2}$), ρ_{f} is the scattering length density of the filler phase ($15.26 \times 10^{10}\,\mathrm{cm}^{-2}$ for carbon black) and ϕ_{r} and ϕ_{f} are the volume fractions of the rubber and filler phase, respectively.

Initially, USAXS experiments were performed at BW4 at detector distances of 8000 and 14000 mm ($\lambda = 0.138$ nm) with a MAR 165 CCD detector. However, it turned out that due to the high scattering power of the carbon black and due to the limited dynamic range of the detector, the absolute scattering intensities were not reliable. Attempts with limited success were made to extend the dynamic range by multiple exposures with different exposure times and merge the scattering patterns during postprocessing. Consequently, all results shown are based on data obtained at MiNaXS with a Pilatus 1 M detector, which boasts a much better dynamic range. The detector distance was 4920 mm, the wavelength 0.1069 nm.

[5] In preliminary experiments, scattering patterns acquired at two different detector distances were merged to extend the region of scattering vectors, but this method was not used for the final data evaluation due to the insensitivity of the change in Q with respect to the integration limits and due to the considerable experimental effort of the method.

3.3 Mechanical Testing

Mechanical tensile tests in the laboratory were performed on an Instron Electropuls E 1000 machine, which is capable of performing quasistatic tests as well as dynamic tests (Fig. 3.14a). Clamps were specifically designed to allow for a secure clamping of the beaded elastomer samples and to reduce sample slippage (Fig. 3.14b).

The strain was measured optically whenever possible. For uniaxial tests, two markers were used. To ensure contrast to the sample, white markers were used for carbon black-filled samples and black markers for all other samples. The markers were applied with white paint, an in-house-optimized mixture of latex emulsion and airbrush paint (white or black). Evaluation was done either fully or partly automatized in PV-Wave [15], using self-written code. The software loads the images and stores the marker position. The marker positions are determined manually by mouse-clicks in the semi-automatic version, which is preferred when the image quality is poor. Provided the photo quality is sufficient, a fully automatized procedure can extract the optical strain data without further user interaction.

In case of a non-homogeneous strain field in the region of interest (e.g. around a crack tip), a speckle pattern was applied to the sample surface by airbrush, and the digital image correlation software GOM Aramis [16] was used. Depending on the requirements, the standard cameras (provided with the Aramis system) were

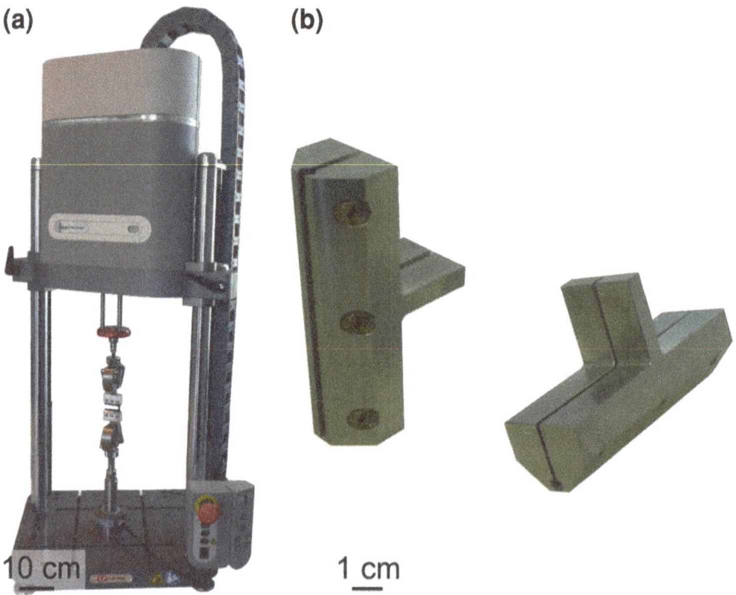

Fig. 3.14 **a** Instron Electroplus E 1000 electrodynamic tensile machine; **b** grips for beaded samples, length 80 mm

employed or photos were imported to Aramis. Postprocessing of the data was done in PV-Wave.

Synchronization between optical strain data and mechanical data is ensured by a trigger signal, issued by the E 1000 tensile machine and passed to the Aramis controller via a cable.

3.4 Optical Volumetric Studies

While USAXS allows to study void formation with good spatial and time resolution, it is no absolute method. It relies on models and can only detect cavities within a certain size range. As outlined in Sect. 1.4, optical methods gained popularity over the last years due to improved digital camera technology. To circumvent the problems associated with digital image correlation at high strains, the basic idea was to take advantage of a rotation symmetric sample, such that from the contour alone one could calculate the volume by dissecting the sample into slices of one pixel height, assuming a circular cross section at any time.

Several types of samples were used. The most simple setup is to glue (or vulcanize) a cylindrical rubber sample between two metal cylinders. While this method works nicely for small strains, the tests cannot be performed up to sample failure because the interface between sample and metal is always the weakest point. To reach larger strains, a sample with dumbbell geometry was designed (Fig. 3.3d). To avoid the formation of pores during vulcanization, it was vulcanized within a heated tablet press, such that a higher internal pressure in the mold cavity was reached than in a conventional plate hot press.

For cylindrical specimens, the complete sample volume was monitored during the tensile experiment, using a 4 Megapixel camera of the Aramis system. Images were taken in small intervals to ensure small changes between consecutive images and facilitate recognition of the contours. The colors of the background and the metal parts were varied to increase the contrast with the sample, i.e. white color was used for carbon black-filled rubbers and black color for calcium carbonate-filled rubbers. The contours between sample and air as well as between sample and metal holders were detected automatically by a self-written procedure in PV-Wave. Edge-enhancement functions in analogy to the shadow detection in Fig. 3.11 were used. An example is shown in Fig. 3.15.

For dumbbell samples, markers were applied to the sample to define a volume to be monitored, since the rest of the sample extends into the clamps. The algorithm automatically followed the position of these markers (dots of white or black latex paint).

The biggest challenge turned out to be the glare patches on the surface of the rubber sample, caused by reflection of the light that was used to illuminate the sample. Turning off the light is no option since this exacerbates the contrast between sample and background. The only solution is to use polarizers for the camera and for the lights.

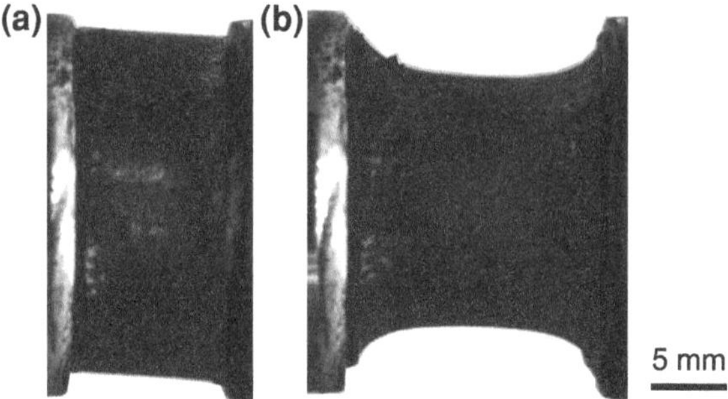

Fig. 3.15 Photographs taken during volumetric experiments. The *dark red area*, representing the detected sample, was automatically generated by a PV-Wave procedure. **a** Undeformed sample; **b** same sample in stretched condition

3.5 Scanning Electron Microscopy

In-situ stretching experiments are significantly more instructive than SEM analysis of ex-situ stretched samples, because one avoids relaxation and can follow structural changes at a certain sample position in real time. Since commercial miniature tensile devices to be used in a SEM can not provide the displacement required for the study of elastomers or exceed the mass limit of 500 g to be mounted on the sample stage of the Zeiss Gemini Ultra Plus, IPF Forschungstechnik designed and produced a machine (Fig. 3.16a) fulfilling the specific requirements (maximum load 50 N, maximum displacement 50 mm). A small control box is used for manual operation of the machine. For the sake of simplicity and lightweight, no load cell is used. To observe the crack from different angles, the sample is rotatable around the tensile axis. An adapter with vacuum-specified connectors was permanently mounted to the SEM sample chamber to run the machine in the chamber. In order to study non-conductive samples by SEM, they usually need to be coated with carbon or a metal (e.g. platinum) to increase their electric conductivity and avoid charge build-up. However, these coating layers are brittle and cannot withstand the high strains of elastomers. Thus, only inherently conductive materials can be studied with in-situ stretching and the recipes need to be optimized to increase the conductivity. High loadings of carbon black were used. To maintain the processability on an acceptable level, large amounts of stearic acid were added as processing oil. In general, finer blacks give better conductivity. However, these are harder to analyze at a given magnification in the SEM. The conductivity of different carbon black grades was compared to obtain a good trade-off between conductivity and particle size (Fig. A.1).

With electron microscopy being a routinely applied method in polymer science, the reader is referred to textbooks for an in-depth explanation of the method [17–19].

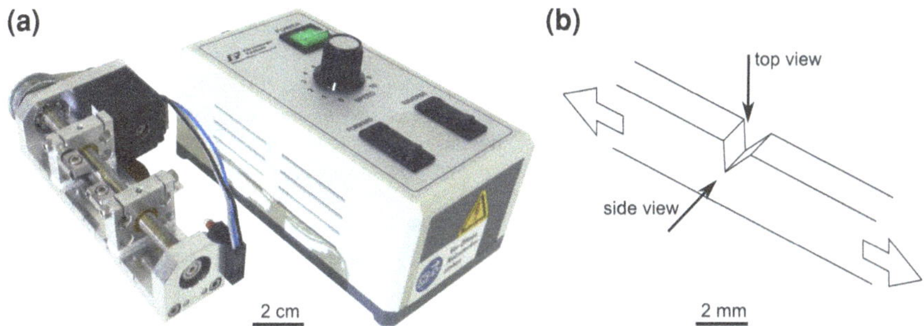

Fig. 3.16 **a** Microtensile machine for in-situ experiments in a SEM. **b** Illustration of *top view* and *side view* of a notched tensile sample, realized by rotation of the sample around the tensile axis

References

1. N. Stribeck, *X-ray Scattering of Soft Matter* (Springer, Heidelberg, 2007)
2. K. Brüning et al., Strain-induced crystallization around a crack tip in natural rubber under dynamic load. Polymer **54**(22), 6200–6205 (2013)
3. P. Rublon et al., In situ synchrotron wide-angle X-ray diffraction investigation of fatigue cracks in natural rubber. J. Synchrotron Radiat. **20**(1), 105–109 (2012)
4. K. Brüning et al., Kinetics of strain-induced crystallization in natural rubber studied by WAXD: dynamic and impact tensile experiments. Macromolecules **45**(19), 7914–7919 (2012)
5. A. Immirzi et al., Crystal structure and melting entropy of natural rubber. Macromolecules **38**(4), 1223–1231 (2005)
6. K. Brüning, K. Schneider, G. Heinrich, Deformation and orientation in filled rubbers on the nano- and microscale studied by X-ray scattering. J. Polym. Sci. Part B: Polym. Phys. **50**(24), 1728–1732 (2012)
7. I. Morfin et al., ASAXS, SAXS and SANS investigations of vulcanized elastomers filled with carbon black. J. Synchrotron Radiat. **13**(6), 445–452 (2006)
8. A. Saupe, W. Maier, Methoden Zur Bestimmung des Ordnungsgrades nematischer kristallinflüssiger Schichten. Der Ordnungsgrad von Azoxyanisol. In: Zeitschrift Naturforschung Teil A **16**, 816 (1961)
9. P.G. De Gennes, *The Physics of Liquid Crystals* (Clarendon Press, Oxford, 1979)
10. B.J. Lemaire et al., The measurement by SAXS of the nematic order parameter of laponite gels. Europhys. Lett. **59**, 55–61 (2002)
11. P. Lindner, T. Zemb, *Neutrons, X-rays, and Light: Scattering Methods Applied to Soft Condensed Matter* (Elsevier, North-Holland, 2002)
12. J.B. Hayter, J. Penfold, Use of viscous shear alignment to study anisotropic micellar structure by small-angle neutron scattering. J. Phys. Chem. **88**(20), 4589–4593 (1984)
13. H. Zhang et al., Nanocavitation in carbon black filled Styrene-Butadiene rubber under tension detected by real time small angle X-ray scattering. Macromolecules **45**(3), 1529–1543 (2012)
14. A. Jánosi, Röntgenkleinwinkelstreuung von Mehrphasensystemen, Interpretation der Elektronendichtefluktuation: II. Dreiphasensysteme. In: Monatshefte für Chemie/Chemical Monthly **114**(4), 383–387 (1983)
15. Rogue Wave Software GmbH, Robert-Bosch-Straße 5, Haus c, 63303 Dreieich, Germany, www.roguewave.com

16. Gesellschaft für optische Messtechnik mbH, Mittelweg 7–8, 38106 Braunschweig, Germany, www.gom.com
17. G. Michler, *Electron Microscopy of Polymers* (Springer, Berlin, 2008)
18. S.L. Flegler, J.W. Heckman Jr, K.L. Klomparens, *Scanning and Transmission Electron Microscopy: An Introduction* (Oxford University Press, New York, 1993)
19. D.B. Williams, C.B. Carter, *Transmission Electron Microscopy: A Textbook for Materials Science* (Springer, New York, 2009)

Chapter 4
Results

The results section of this work does not intend to give an exhaustive presentation of all results obtained over the last 4 years. Instead, the focus will be on particularly instructive or novel findings and methodological advancements. The majority of the presented results have been published at various conferences and in peer-reviewed journals (see List of Publications in appendix).

4.1 Crystallization Under Quasistatic Load

4.1.1 General Structure-Property Relationships

4.1.1.1 Filler Content

As is well known from literature, the presence of reinforcing fillers shifts the onset of SIC to lower strains (Figs. 4.1 and 4.2) [1]. This was confirmed for a series of NR compounds with 20, 40 and 60 phr of N234 carbon black (Fig. 4.1). Even though crystallization sets in at lower strains, the crystallinity at a given strain can be lower in filled rubbers as compared to unfilled rubbers, because the increase in crystallinity with strain is steepest in unfilled rubbers. This behavior can be ascribed to the inhomogeneous strain distribution in filled rubbers. Due to the rigidity of the fillers, the local deformation becomes non-affine and heterogeneous, such that due to the local confinement, the strain is locally sufficiently large to allow for the onset of crystallization, while the external strain is still low. While all available literature agrees that the crystallization onset is shifted to smaller strain in filled rubbers, the influence of filler content on the slope of the crystallinity versus strain plot is less clear. Unaffected slopes were reported in Refs. [2, 3], slightly reduced slopes were reported in Refs. [4, 5] and a significant reduction in slope, comparable to Fig. 4.1 was reported in Ref. [6].

K. Brüning, *In-situ Structure Characterization of Elastomers during Deformation and Fracture*, Springer Theses, DOI: 10.1007/978-3-319-06907-4_4,

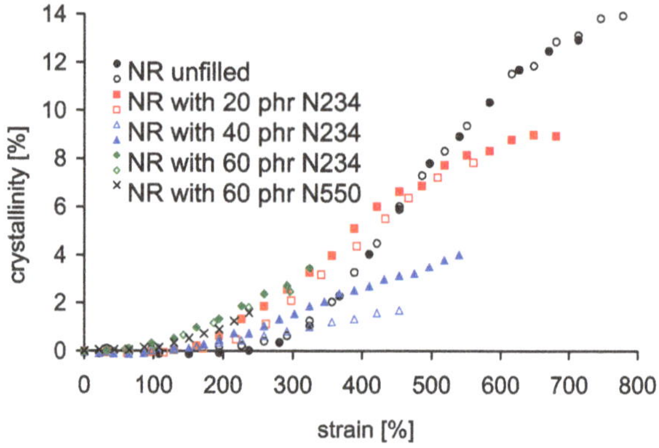

Fig. 4.1 Crystallinity versus strain for NR samples with various filler contents [#2, #3, #4, #5, #6]. For all samples (except NR with 60 phr N550), two data series from different beamtimes with different detector setups are displayed

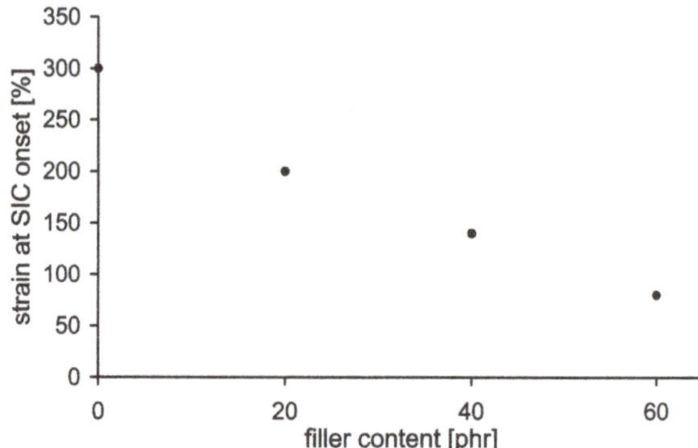

Fig. 4.2 Onset of strain-induced crystallization versus filler content for the sample series containing N234 carbon black (unfilled [#2], 20 phr [#3], 40 phr [#4], 60 phr [#5])

4.1.1.2 Pure Shear Versus Simple Tension Experiments

The degree of crystallinity under pure shear or simple tensile loads was found to follow the same dependency on strain. The crystallization mechanism appears to be unaffected by the confinement of the lateral contraction to one dimension, which is essentially the difference between simple tensile and pure shear loading. This is exemplified in Fig. 4.3. Usually, the maximum degree of crystallinity in pure shear samples is below 5 %, because the elongation at break is significantly reduced compared to simple tensile samples. The results are in agreement with Beurrot et al. [7]. In contrast to this, Nyburg reported a higher crystallinity in pure shear versus simple tension geometry at a given strain [8].

4.1.1.3 Crystallization Under Repeated Loading

The crystallization process itself is well reproducible and insensitive to previous loading histories. As shown in Fig. 4.4b, there is some hysteresis between the loading and unloading branch, but the crystallinity curves for any of the loading cycles coincide.

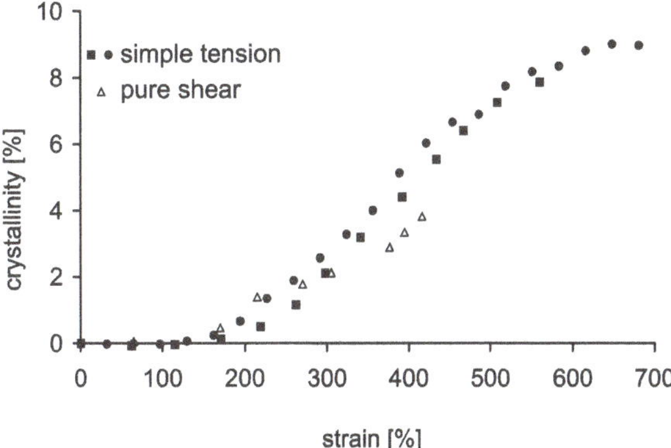

Fig. 4.3 Comparison of strain-induced crystallization in pure shear and simple tension geometries (Fig. 3.3). NR with 20 phr N234 [#3]

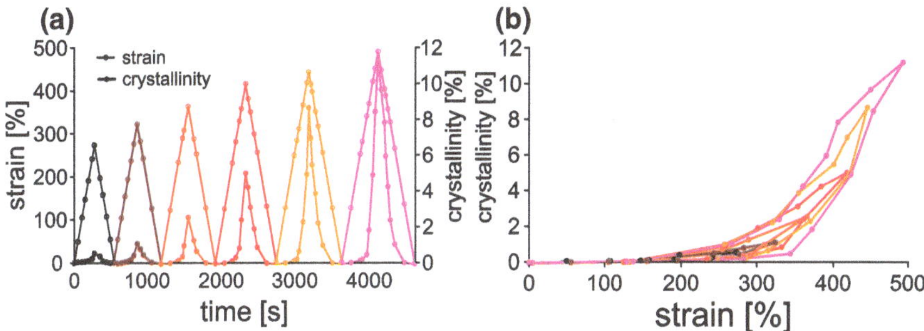

Fig. 4.4 Strain-induced crystallization during repeated loading in unfilled NR [#2]. **a** Strain history over time and resulting crystallinity. **b** Crystallinity versus strain

When performing the same experiment on carbon black filled rubber, the structural results clearly reflect the Mullins effect (Fig. 4.5). When loading the sample repeatedly, larger strains are required to obtain the same degree of crystallinity (Fig. 4.6a). The author hypothesizes that local overstrains are reduced from cycle to cycle due to the breakdown of filler clusters. A plot of crystallinity versus stress (Fig. 4.6b) shows that the increase in strain at a given crystallinity, as might be inferred from Fig. 4.6a, is even overcompensated by the Mullins effect, such that in any consecutive cycle, the stress level at a given crystallinity is reduced compared to the previous cycle. In analogy to the mechanical consequences of the Mullins effect, the structural properties settle after a few cycles (Sect. 4.2.2). Trabelsi reported all cycles of a repeatedly stretched filled sample to be the same in terms of crystallinity, despite the presence of the Mullins effect in the stress-strain behavior [3]. On the contrary, Chenal et al. found an influence of the Mullins effect on the crystallization behavior of repeatedly loaded samples, in analogy to Fig. 4.6 [9].

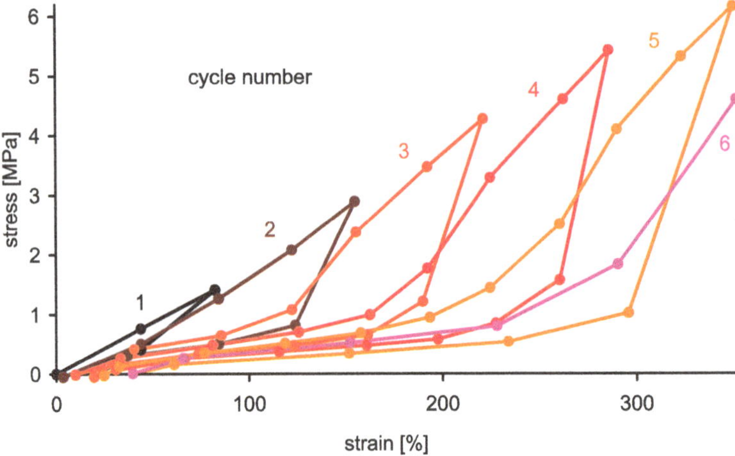

Fig. 4.5 Stress-strain curve of Mullins test performed on NR filled with 60 phr N234 carbon black [#5]. (The small number of datapoints is owed to the fact that the optical strain was evaluated only during hold periods to acquire a WAXD pattern.)

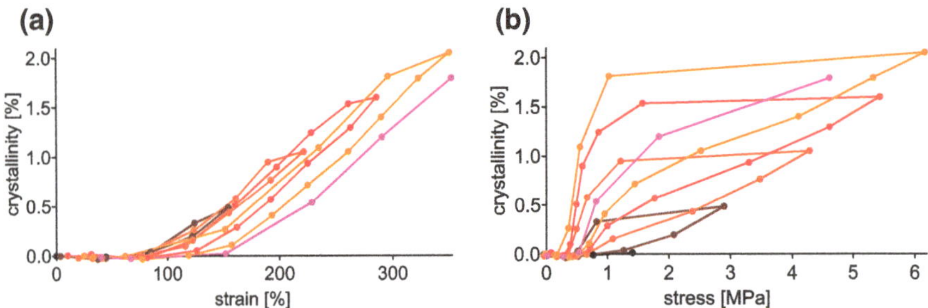

Fig. 4.6 Strain-induced crystallization during Mullins experiment in NR filled with 60 phr N234 carbon black [#5]. The Mullins effect results in an evolution of crystallinity over the number of loading cycles. **a** crystallinity versus strain, **b** crystallinity versus stress

4.1.1.4 Effect of Mechanical Aging on Strain-Induced Crystallization

In order to mimic the loading conditions in real-life applications or in a tear fatigue experiment, tensile samples were subjected to repeated loading at 1 Hz for approximately 1×10^5 cycles before being placed in the beamline. The amplitude was chosen to be as large as possible, such that the sample would not fail before reaching 1×10^5 cycles. In none of the materials was an effect of mechanical aging on the crystallization behavior observed, as exemplified in Fig. 4.7. The role of potential material relaxation between ex-situ aging and the in-situ tensile experiments was not investigated. Beurrot et al. performed in-situ fatigue experiments and found an evolution (increase or decrease, depending on fatigue conditions) of crystallinity over the number of fatigue cycles [10].

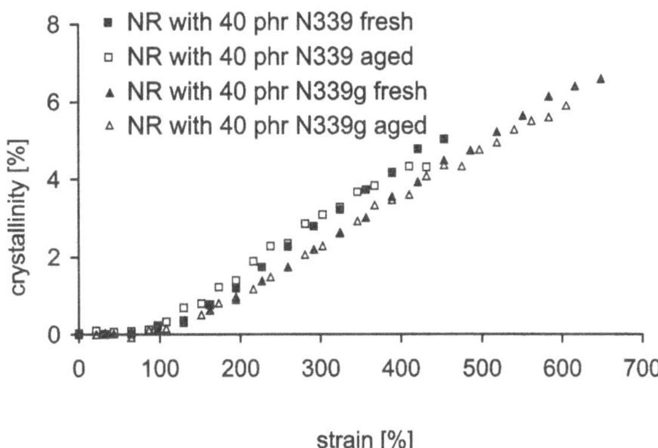

Fig. 4.7 Comparison of SIC in mechanically aged and fresh samples [#7, #8]. *g* stands for graphitized N339 carbon black. (For details regarding the graphitization, see Klüppel [11].)

4.1.1.5 Strain in Crystallites

Deformations in the crystalline structure are reflected in a broadening and shifting of the diffraction peaks. Under strain, a shift of the (120) and (200) peaks towards larger scattering vector magnitudes q by approximately $0.03\,\text{nm}^{-1}$ is observed. At the same time, the (201) peak shifts to smaller angles. The corresponding changes in the unit cell vector magnitudes, $|\boldsymbol{a}|$, $|\boldsymbol{b}|$ and $|\boldsymbol{c}|$ were calculated as follows, assuming an orthorhombic lattice [12]:

$$\begin{aligned} |\boldsymbol{a}| &= \frac{4\pi}{q_{200}} \\ |\boldsymbol{b}| &= \sqrt{\frac{4}{\left(\frac{q_{120}}{2\pi}\right)^2 - \frac{1}{|\boldsymbol{a}|^2}}} \\ |\boldsymbol{c}| &= \frac{2\pi}{\{q_{201}\}_c}. \end{aligned} \tag{4.1}$$

$\{q_{201}\}_c$ is the projection of the (201) peak onto the c-axis. The obtained unit cell dimensions are plotted in Fig. 4.8. The maximum strain in the crystallites along the c-direction is around 0.6 %, while in the transversal a and b-directions, the contraction amounts to roughly 0.8 % (Fig. 4.9). As demonstrated in Fig. 4.8b, the curves for different rubbers coincide when plotted as a function of stress. Approximating the deformation behavior of the crystallites as linear elastic, and assuming that crystallites and amorphous phase are in series, i.e. the crystalline phase bears all the load, the resulting modulus of the crystallites is 3 to 4 orders of magnitude larger than the small-strain tensile modulus of the amorphous phase. Thus, the common approximation of a rigid crystalline phase is well justified. These results are qualitatively in line with literature [2, 13, 14].

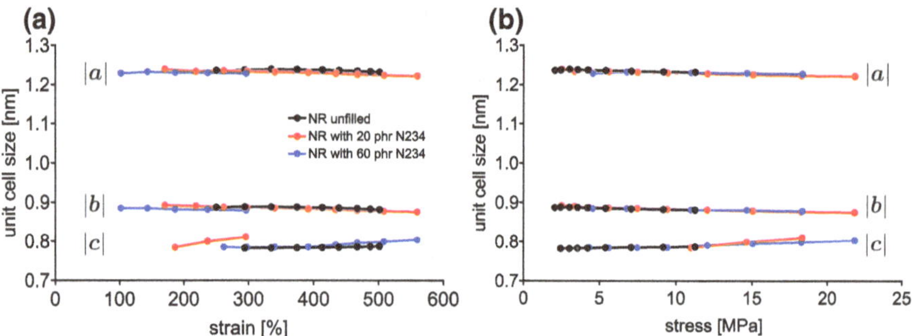

Fig. 4.8 Unit cell vector magnitudes $|\boldsymbol{a}|$, $|\boldsymbol{b}|$ and $|\boldsymbol{c}|$ as a function of **a** strain, **b** stress. Unfilled [#1] and filled NR [#3, #5]. (The graphs intend to show the variation of cell dimensions; the absolute values might deviate slightly from literature due to inaccuracies in the calibration of the detector setup.)

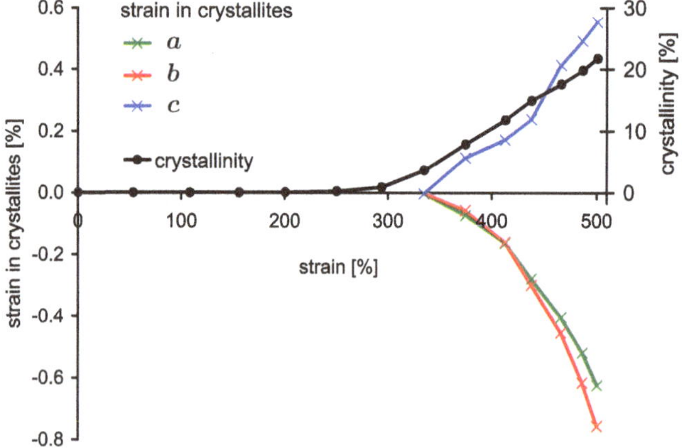

Fig. 4.9 Strain along the crystal axes $\boldsymbol{a}$, $\boldsymbol{b}$ and $\boldsymbol{c}$ as a function of global optical strain in unfilled rubber [#1], along with crystallinity. For low crystallinity values, the deformation of the unit cell could not be quantified and the values at 335 % external strain were taken as reference

4.1.2 Strain Amplification in Filled Rubbers

As outlined in Sect. 1.1.4, the concept of strain amplification serves to study the micromechanical behavior of the soft phase in a composite system. The degree of crystallinity Φ was employed as a quantity to derive the local strain state in the matrix of a filled elastomer. The underlying assumption is that a unique relation exists between the degree of crystallinity and the strain, and that this relation holds in an unfilled material as well as in the matrix of a filled material. Then one can calculate the strain amplification factor A_Φ from Eq. 1.8. The results are shown in Fig. 4.10. It is clear that a reinforcing filler, such as carbon black, in general has a strain amplification factor A_Φ greater than unity, as expected from theory. However,

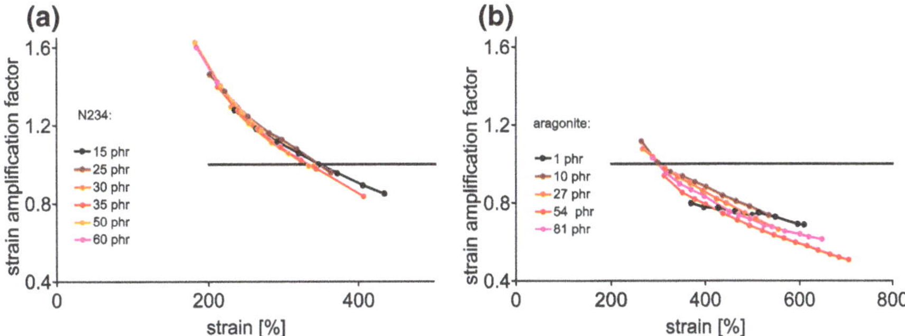

Fig. 4.10 Strain amplification factor A_Φ derived from strain-induced crystallization. **a** carbon black (N234) filled NR [#10, #12, #13, #14, #17, #19], **b** aragonite filled NR [#20, #21, #22, #23, #24]. The horizontal line at $A_\Phi = 1.0$ is for orientation. Aragonite loadings of 27, 54 and 81 phr correspond to carbon black loadings of 20, 40 and 60 phr in terms of volume fraction

A_Φ is not strain-independent. In general, it decreases with strain. This can be ascribed to the method, taking the degree of crystallinity as relevant quantity. At high degrees of crystallinity, the confinement of chains due to the presence of carbon black is assumed to hinder the crystallization [3]. What is more interesting is that there is no clear dependency of A_Φ on the filler content. This becomes even clearer when drawing the comparison between carbon black filled materials and aragonite filled ones. Aragonite, a rod-shaped form of precipitated calcium carbonate (Fig. 3.13), was chosen as a non-active filler. According to the simplified hydrodynamic strain amplification model, only the volume content of the filler governs the matrix overstrain, as long as no cavitation occurs. Clearly, in the aragonite filled samples, crystallization is shifted to larger strains by the sheer presence of aragonite, as can be seen from the 1 phr-sample, where the low filler loading can be assumed to have a negligible effect on the matrix strain. This shows that the degree of crystallinity should be used with caution to derive the matrix strain. We hypothesize that the crosslinking density is affected by the aragonite due to adsorption of crosslinking agents on the filler surface. To a lesser extent, this is also observed in carbon black filled rubbers, particularly when filled with highly active carbon black grades, but most often it is overlooked in the discussion of strain amplification [15].

Disparity between strain amplification factors derived from different quantities is generally reported in literature [3, 9]. Trabelsi et al. also found a decrease of A_Φ with strain [3].

Thus, we can make the conclusion that hydrodynamic strain amplification might be one of the factors contributing to reinforcement, but it is certainly not the major one.

4.1.3 Local Structure Around a Crack Tip

In the context of tear fatigue, the local structure in the vicinity of the crack tip governs the crack propagation behavior. Scanning WAXD with a microbeam is the method of choice to study strain-induced crystallization around the crack tip with a spatial resolution in the micron range. Owing to the overstrain in the vicinity of the crack tip, the material starts to crystallize at low external strains. This is especially pronounced in filled rubbers. In notched[1] samples of pure shear geometry (Fig. 3.3c), first signs of crystallinity were observed at external strains as low as 20 % in samples with a high carbon black content (not shown). The crystalline area does not extend into the sample by more than 1 mm (at large strains right before catastrophic crack growth, Figs. 4.11 and 4.12). In filled samples, the crystalline area is much larger than in unfilled, contributing to a more pronounced self-reinforcement and better tear fatigue properties.

In contrast to pure shear samples, notched tensile sample (SENT) can become completely crystalline before failure occurs. The results are exemplified for natural rubber with 60 phr N234 in Fig. 4.13. The regions subjected to the largest strains also exhibit the highest degree of crystallinity. A further refinement in spatial resolution down to 30 μm revealed a narrow region directly below the crack surface with substantially higher crystallinity. As expected, the crystallites around the crack tip are oriented along the principal strain direction (Fig. 4.14).

The results are in good qualitative agreement with literature [16–18]. A quantitative comparison is not possible due to different sample geometries (e. g. crack length).

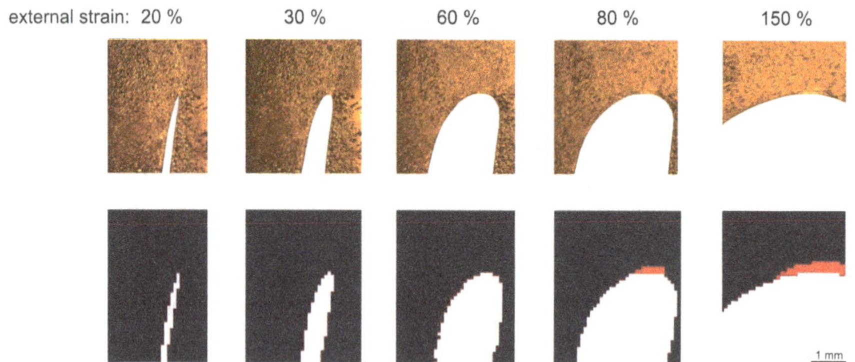

Fig. 4.11 Crack tip scans of a notched pure shear sample of unfilled NR [#1] at different strain steps. The *top row* shows photographs of the sample. The *bottom row* shows crystallinity maps, with the crystalline zone marked *red*. The *scale bar* represents 1 mm. As described in Sect. 3.2.6, *each pixel* represents 100 × 100 μm

[1] During sample preparation, the notched samples were manually torn to propagate the crack a little bit and to convert the notch tip into a sharp crack tip.

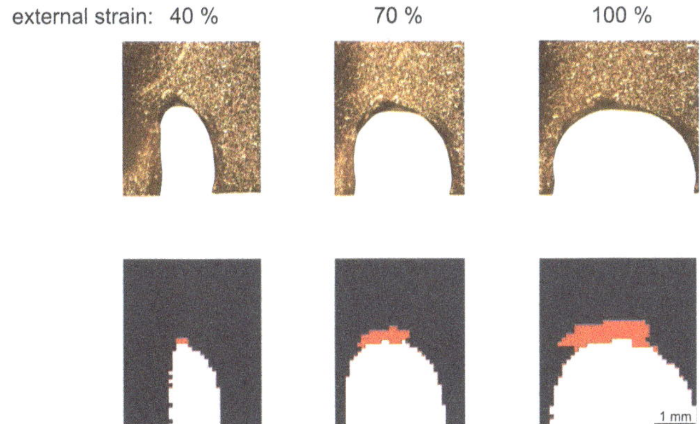

Fig. 4.12 Crack tip scans of a notched pure shear sample of NR with 20 phr N234 [#11] at different strain steps. The crystalline zone (*red*) is much larger than in unfilled NR

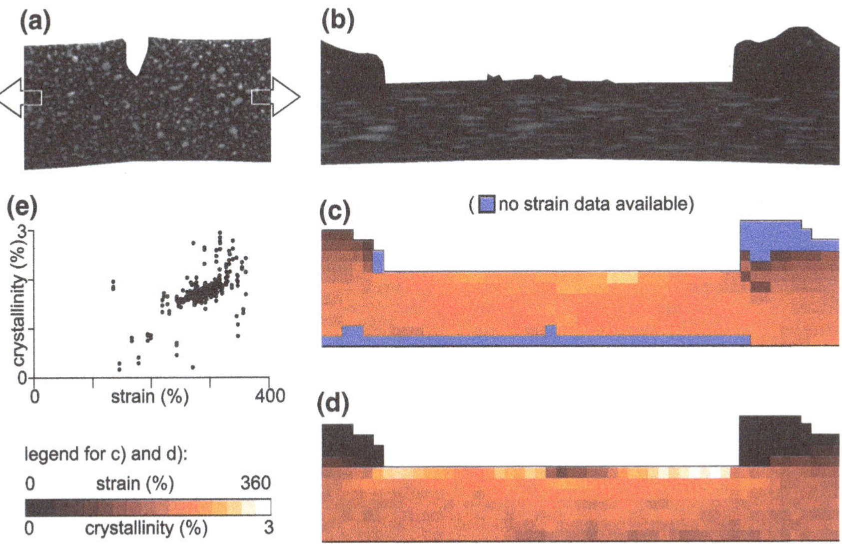

Fig. 4.13 Crack tip scan of SENT specimen of natural rubber with 60 phr N234 [#25] at an external strain of 250 %. **a** Photograph of the sample under very moderate strain. The speckle pattern serves for the strain field evaluation by digital image correlation. *Arrows* indicate the stretching direction. **b** Photograph of sample under high strain, the crack has propagated. **c** Strain field analysis (strain along the tensile direction). **d** Crystallinity obtained by scanning WAXD. *Each pixel* in (**c**) and (**d**) represents $100 \times 100\,\mu$m. **e** Plot of crystallinity over strain extracted from the scan data. The relation is similar to the simple tensile case

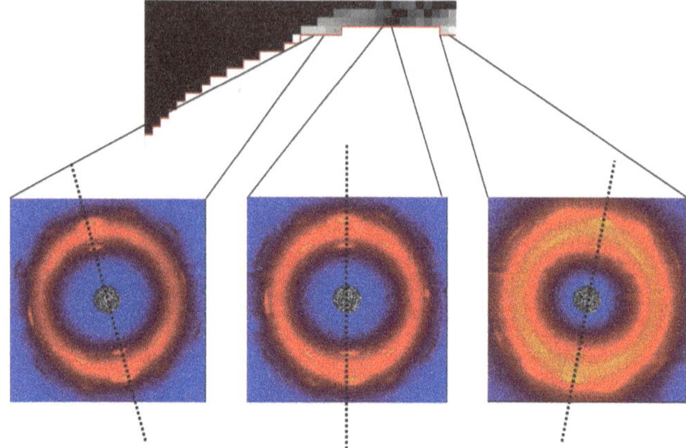

Fig. 4.14 Orientation of the crystallites around a crack tip, exemplified on unfilled NR [#1] (cf. Fig. 4.11 right). The *greyscale* represents the orientation. Three WAXD patterns from representative locations are shown

4.1.4 In-situ Observation of Static Crack Growth and Deflection

Crack deflection and branching, resulting in a wrinkled crack propagation pathway, are characteristic for natural rubber and have been ascribed to SIC [19, 20]. Crack deflection and branching on various length scales dissipate a large amount of strain energy and contribute to the tear fatigue resistance of NR. In this section, the structural fundamentals behind crack deflection and branching are demonstrated.

The notched SENT sample was stretched quasistatically and static two-dimensional WAXD scans were performed at certain intervals, when the crack propagation mechanism changed. Initially, the stress and strain have their maxima at the notch tip, accompanied by the maximum crystallinity in this region (Fig. 4.15a). This has frequently been described in literature. However, the crack did not propagate through this region of high strain, because its tensile strength is elevated by the highly oriented crystallites. Instead, after increasing the external strain, the crack branched and propagated sideways, close to the boundary of the crystalline zone, through a region of lower crystallinity. Here, despite the increase in sample cross section, the high gradient in crystallinity leads to a local stress concentration with respect to the tensile strength. Moreover, the crystallites reinforce the material mainly in the tensile direction, which additionally favors crack deflection. Following the crack propagation, the maxima in strain and crystallinity are now located around the new crack tips (arrows in Fig. 4.15b, c). The relation between crystallinity and strain, extracted from the crystallinity and strain maps after coordinate transformation, is shown in Fig. 4.16.

The observed mechanisms of crack branching under static conditions are assumed to apply to dynamic crack growth as well. Once the crack has grown sideways for a couple of cycles, it is deflected again, due to similar reasons as before, and

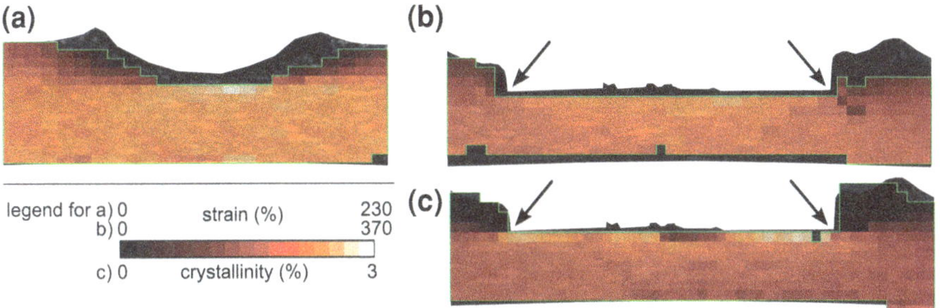

Fig. 4.15 Scanning WAXD performed on a strained SENT specimen. The sample consists of carbon black reinforced (60 phr N234) natural rubber [#25]. **a** Strain field (strain in stretching direction) obtained from digital image correlation superimposed on a photograph of the sample under 140 % external strain. Stretching direction is horizontal. **b** Photo with strain field of sample under 230 % external strain, the crack has propagated sideways. **c** Crystallinity obtained by scanning WAXD, sample is in same state as in (**b**). *Each pixel* represents $100 \times 100\,\mu m$

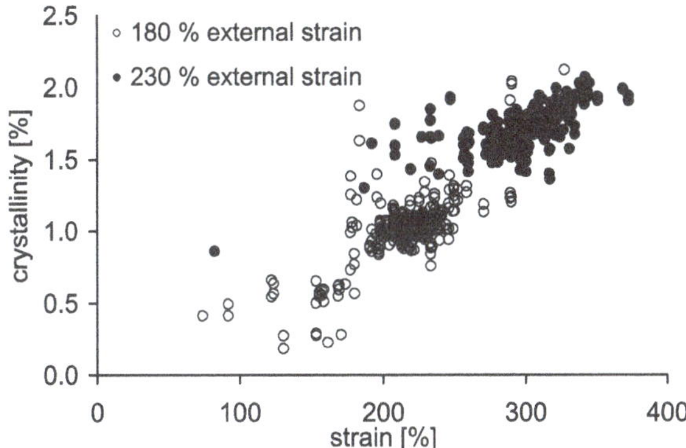

Fig. 4.16 Crystallinity versus strain obtained from two crack tip scans at external strains of 180 and 230 % (cf. Fig. 4.15)

additionally the external strain favors a propagation perpendicular to the stretching direction. Consequently, the crack propagates again in the initial direction. This behavior is demonstrated in Fig. 4.17. Unfortunately, it is currently impossible to make 2D scans of a propagating crack.

4.1.5 Crystallization Under Biaxial Load

As compared to classic uniaxial testing, multiaxial strain more closely resembles the loading conditions in real rubber parts. By changing the ratio between the displacements of the two axes, one can gradually move from uniaxial strain to equibiaxial strain. When applying an equibiaxial displacement to the cruciform specimen

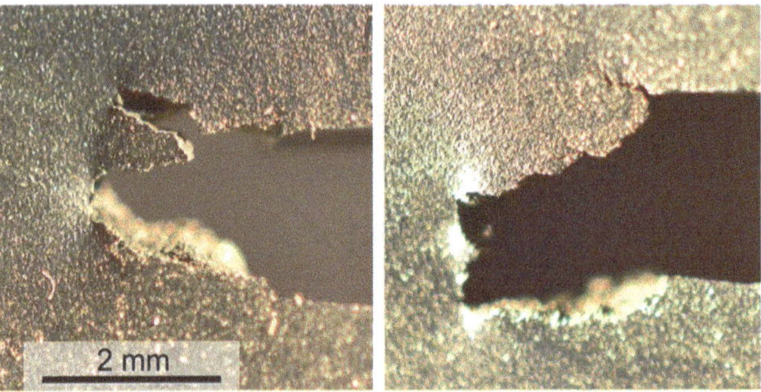

Fig. 4.17 *Photographs* showing crack branching. The samples (pure shear geometry, NR with 60 phr carbon black [#25]) were cycled for 8000 cycles with an amplitude of 30 % strain ($R = 0$) with a frequency of 1 Hz (*sinusoidal*). The scale bar represents 2 mm and applies to both images

(Fig. 3.3b), the sample undergoes SIC starting in the arms and along the edges. In these regions, uniaxial strain prevails and thus fiber symmetry (with a varying orientation of the fiber axis) was assumed. All crystallites are highly oriented along the principal strain directions. One quadrant of the sample stretched by 35 mm along both axes is shown in Fig. 4.18. The figure was obtained by scanning WAXD with a microbeam of $35 \times 25\,\mu$m and a step width of $500\,\mu$m in both directions, i.e. each pixel in Fig. 4.18 represents an area of $500 \times 500\,\mu$m in real space. For each of the 999 WAXD patterns, the crystallinity and orientation were computed. Considering that during one scan the material is exposed to the beam for almost 10 min, larger elongations than 35 mm lead to sample failure during the scan due to beam damage and thus no scans of highly biaxially stretched samples could be obtained. The central part of the cruciform specimen remained amorphous. The maximum principal strain in the center of the sample was below the uniaxial SIC onset strain. Using

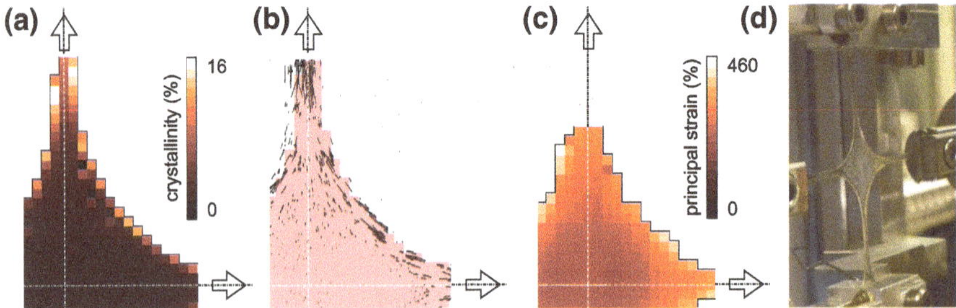

Fig. 4.18 Crystallinity in an equibiaxially stretched unfilled NR [#1] sample. One quadrant of the cruciform sample is shown. *Horizontal* and *vertical* displacements are 35 mm each. *One pixel* represents $500 \times 500\,\mu$m. **a** Degree of crystallinity, **b** Orientation of the crystallites (the length of the arrows is proportional to the crystallinity), **c** Magnitude of principal strain, determined by ex-situ digital image correlation. The values are around 220 % in the biaxially loaded center and around 400 % along the edges, **d** Photograph of the highly stretched sample

a different sample geometry, Beurrot et al. observed crystallinity under equibiaxial strain at room temperature. However, no attempt to quantify the crystallinity was made [7].

With the help of combined thermal and strain-induced crystallization, it was shown in the literature, that under biaxial strain the crystallites orient with their *b*-axis normal to the sample plane, whereas the *a*- and *c*-axis are oriented randomly [21–23]. This implies that of course the fiber symmetry no longer holds and that without further assumptions, the crystallinity can only be quantified after rotation around more than one axis, requiring a sophisticated experimental design. At present, the general relation between crystallinity and non-uniaxial strain fields is unknown.

4.2 Kinetics of Strain-Induced Crystallization in Natural Rubber

Considering that most rubber parts, e.g. tires, are predominantly subjected to dynamic cyclic loads (Fig. 1.20), not only the equilibrium structure, but also the transient structures and the rate of structure formation are important. The mechanical characterization of rubber materials under dynamic conditions is done routinely; however the structural characterization presents a challenge, since it requires experimental techniques with very high time resolution. Owing to the increase in brilliance of third generation synchrotrons by roughly three orders of magnitude over second generation sources, exposure times could be reduced from the second range to the millisecond range over the last 15 years. Along with the sources, the capabilities of X-ray detectors also made significant progress over the last decade. Since 2005, pixel detectors like the Pilatus series developed by the Paul Scherrer Institute (Switzerland) are commercially available and have become more affordable over the years. The Pilatus 300 K detector features a readout time of 3 ms, as opposed to several seconds in classic CCD detectors. The readout process does not add any noise to the signal, such that good signal-to-noise ratios can be obtained at very short exposure times, and multiple frames can be combined without accumulating noise. The MiNaXS beamline at Petra III provides the powerful computer infrastructure, which is required to process data rates as high as 140 MB/s over more than a minute. Taking advantage of all these capabilities, WAXD experiments with a pattern acquisition rate of more than 140 fps were performed, corresponding to exposure times of 4 ms [24].

To tackle the SIC kinetics, two approaches were followed: (a) In analogy to pressure jump or temperature jump techniques applied in the study of chemical reaction kinetics, the governing variable of SIC, the strain, was changed at a rate faster than the crystallization rate, and then the system was observed to return to equilibrium over time at constant strain. (b) In order to mimic real-life loading conditions more closely, the sample was subjected to a harmonic cyclic strain at a frequency of 1 Hz and the crystallization process was followed with a time resolution capable of resolving each cycle individually as well as following the trend over several cycles.

The experiments were made with unfilled NR [#1], filled NR (20 phr N234 carbon black [#11]) and synthetic isoprene rubber (IR) [#26].

4.2.1 Tensile Impact Experiments

Tensile impact experiments were carried out on a specifically designed tensile machine (Fig. 3.10), stretching the tensile sample (Fig. 3.3a) to a predefined strain within less than 10 ms. The synchronization between the tensile machine and the WAXD patterns was ensured by a shadow on the detector, which moved upon the release of the spring.

Provided the strain step was sufficiently large, crystalline diffraction peaks were usually observable on the very first pattern after the strain step, i.e. less than 10 ms after completion of the strain step (Fig. 4.19a). The required strain step heights were 400 % for unfilled NR, 300 % for filled NR and 570 % for IR. Only in a limited region of strains slightly below these values, was crystallization observed to set in after a longer time (Fig. 4.19b). The crystallinity directly after the strain step is already considerable, around half of the final crystallinity Φ_f after 60 s. Long-time experiments show that beyond 60 s, the crystallinity can be considered constant (4.20). After the strain step, crystallization proceeds, following a stretched exponential function, and gradually approaches the final level Φ_f, as shown in Fig. 4.21.

The most popular model for the description of crystallization kinetics is the Avrami model:

$$\Phi_{\text{fit}}(t) = \Phi_f \left(1 - \exp\left(-(kt)^{m+\delta}\right)\right). \tag{4.2}$$

Fitting the experimental data by the Avrami equation yields exponents $m + \delta$ below 0.5, which is incompatible with the physics behind the model. The Avrami model

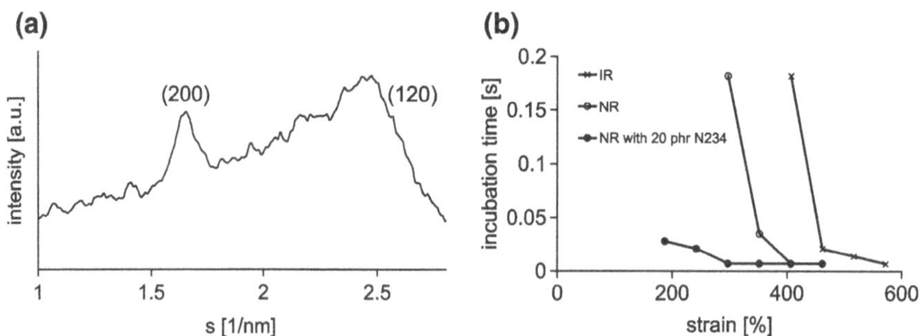

Fig. 4.19 **a** Equatorial diffraction curve of NR, obtained from the first pattern after completion of the strain step. The first signs of (200) and (120) crystalline peaks are clearly visible. **b** Incubation times after strain steps in different materials [#26, #1, #11]. The incubation time is defined as the time period from completion of the strain step to the detection of the first crystalline peaks

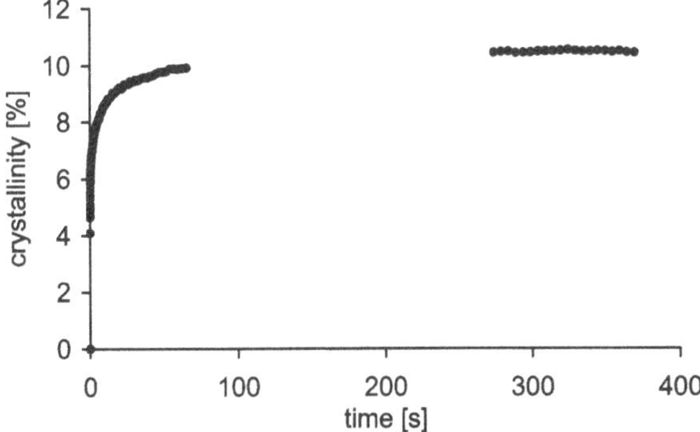

Fig. 4.20 Crystallinity Φ versus time in NR [#1] for a strain step of 410 %. After 60 s, the crystallinity can be considered stable

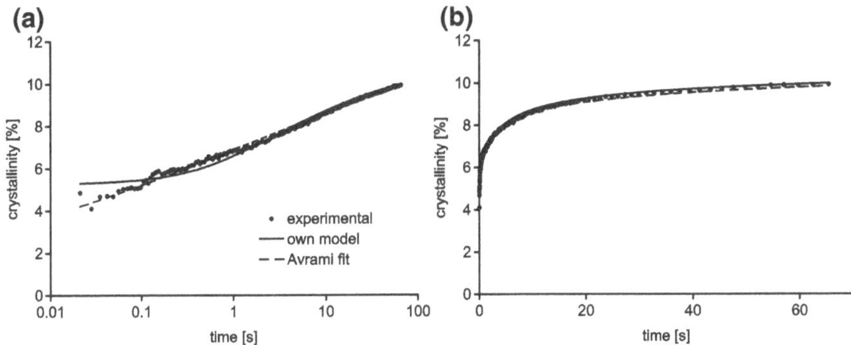

Fig. 4.21 Crystallinity Φ versus time in NR [#1] for a strain step of 410 % with fits acc. to Eq. 4.4 (*solid line*) and Avrami fit (*dashed line*, $m + \delta = 0.18$): **a** logarithmic time scale, **b** linear time scale

postulates a two step process, consisting of nucleation and growth. It predicts an exponent $m + \delta$ for a linear growth of the crystallites in m dimensions, with $\delta = 0$ if the nucleation rate is 0 and $\delta = 1$ if the nucleation rate is constant (and greater than 0) [25]. The mismatch between the Avrami model and the experimental data of the SIC kinetics does not come as a surprise, since the Avrami model was intended for thermal crystallization. In fact, as Mandelkern pointed out in 2004, the current understanding of the SIC mechanism is still in its infancy and a lot of work, both theoretical and experimental, remains to be done [26]. Other points disfavoring the use of the Avrami equation are different morphologies (spherulitic vs. fibrillar), different size scales (crystallite sizes are only about 10 nm in SIC), anisotropy due to the strain (reduced diffusion transversal to the strain) and the presence of crosslinks (rendering large scale diffusion impossible). Other mechanisms suggested in the literature include an instantaneous coil-to-stretch transition, originally put forward

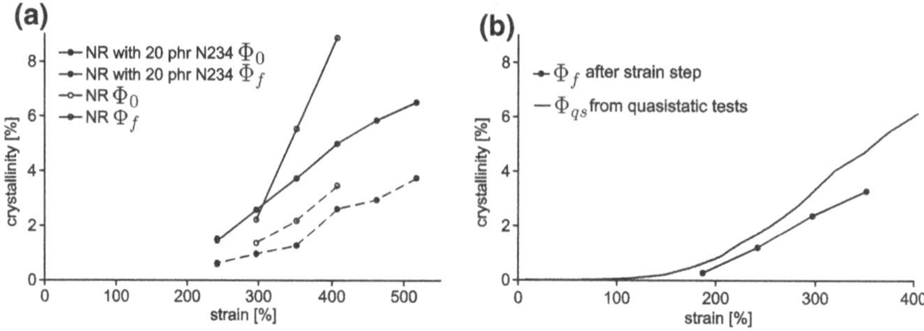

Fig. 4.22 **a** Initial crystallinity Φ_0 (*dashed lines*) and final crystallinity Φ_f (*solid lines*) versus strain step height for NR [#1] (*open symbols*) and filled NR [#11] (*filled symbols*). **b** Final crystallinity after step strain Φ_f in comparison with crystallinity Φ_{qs} obtained from quasistatic experiments in filled NR [#11]

by De Gennes and later modified for the application to networks by Hsiao [27, 28]. He proposed that in networks the transition takes place in chain segments between crosslinks. In fact, the crystallites observed in SIC are smaller than the distance between crosslinks. In line with this model, we postulate that the crystallization mechanism consists of two steps: First an instantaneous nucleation and second a growth phase. The growth rate is proportional to the number of crystallizable chain segments to the power of n, where n describes the propensity of amorphous chains towards crystallization:

$$\frac{d\Phi}{dt} = (k\,(\Phi_f - \Phi))^n \,. \tag{4.3}$$

Integration gives

$$\Phi(t) = \Phi_f - \left((n-1)\,k^n t + (\Phi_f - \Phi_0)^{1-n}\right)^{\frac{1}{1-n}} \,. \tag{4.4}$$

Φ_0 is the crystallinity after the nucleation step,[2] t is the time and k is a proportionality constant. The quality of the fit is demonstrated in Fig. 4.21. The crystallinity after nucleation, Φ_0, and the final crystallinity, Φ_f, are plotted over strain in Fig. 4.22a. The exponent n was found to be close to 4 in all cases. Despite being so far off equilibrium after the strain step, the material approaches almost the same final state as in quasistatic experiments, i.e. the crystallinity measured one minute after the strain step, Φ_f, is close to, but slightly lower than, the crystallinity values obtained from quasistatic tensile tests, Φ_{qs} (Fig. 4.22b).

Application of the Scherrer equation to the diffraction data yields an increase in crystallite size over time in the a and b crystallographic directions, whereas no growth

[2] For the sake of brevity, the crystallinity after the nucleation step is termed *initial crystallinity* in the following. It is a fitting parameter, but closely resembles the first measured crystallinity values for $t \to 0$. Thus its determination presumably depends on the time resolution of the experiment.

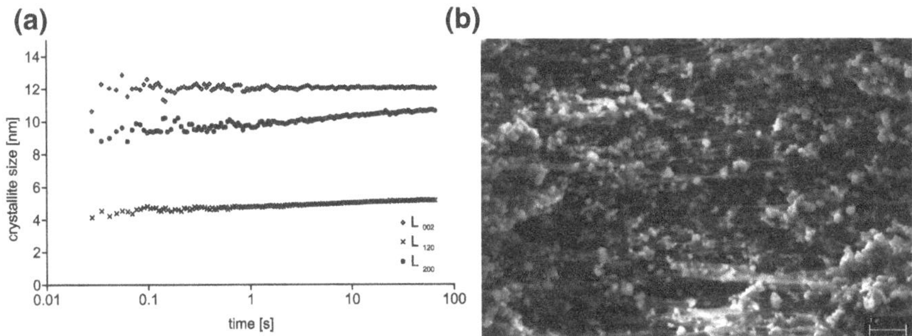

Fig. 4.23 **a** Crystallite size L_{hkl} versus time after a strain step of 410% in NR [#1]. **b** SEM micrograph of highly stretched NR filled with 150 phr N772 carbon black [#30]. The *scale bar* represents 1 μm

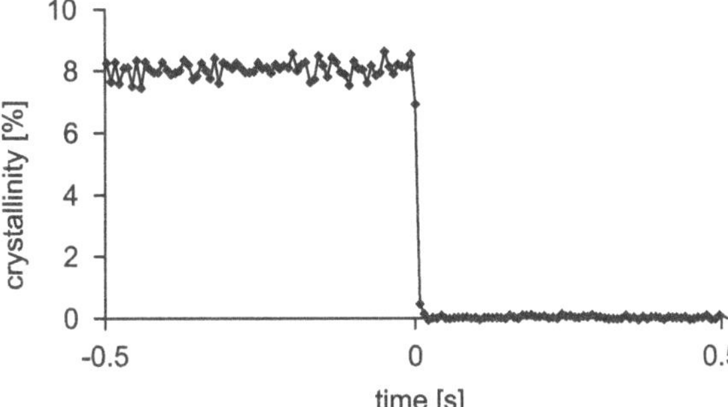

Fig. 4.24 Crystallinity versus time in IR [#26] after steplike unloading from 500 to 0% at $t = 0$ s. The crystallites disappear instantaneously. The time between two consecutive points is 7 ms

takes place in the *c* direction, which corresponds to the tensile direction (Fig. 4.23a). This can be explained by the arrangement of the crystallites. Similar to shish structures formed in flow-induced crystallization, the fibrils in SIC are apparently made up of small individual crystallites of roughly 10 nm in size, which arrange in rows to form the fibrils. This means that growth can only take place in the radial *a* and *b* directions. SEM observations (Fig. 4.23b) suggest that the fibrils are around 20–50 nm in diameter, which is significantly larger than the sizes derived by the Scherrer equation. The reason for the discrepancy could be that either fibrils are made up of several crystallites in their cross section, or the crystalline core is surrounded by oriented amorphous material.

The melting kinetics are much faster than the crystallization kinetics. After steplike unloading (within less than 10 ms) of any of the materials under investigation, the crystalline peaks disappeared instantaneously on the experimental accessible time scale, i.e. within approximately 10 ms. This is exemplified for IR in Fig. 4.24.

The ultra-short incubation times of SIC have so far not been reported in literature. The lack of direct insight into the very early stages of crystallization after a strain step can be assigned to the unavailability of sufficiently fast detectors and stretching machines. Dunning overcame these problems in 1967 by performing WAXD on a continuously stretched rubber band that was passed between two rolls rotating at different angular velocities, stretching the rubber band within a few milliseconds up to 500 % [29]. He found an SIC incubation time of at least 0.1 s and observed that the crystallinity increased constantly over more than one hour. On the contrary, Mitchell used thermal techniques to measure SIC after a strain step and reported that crystallization was almost complete within less than 0.5 s. The incubation time was detected to be well below 0.1 s [30]. Other studies approached SIC indirectly by volume measurements or stress relaxation experiments [31–33]. Most of these studies gave crystallization half times longer than 10 min. The incorrectness of the assumed underlying relationship between stress and crystallinity was later clarified by Yeh [34].

Recently, a direct investigation of tensile impact tests was done by Tosaka et al. [35–37]. In their latest setup, it took around 500 ms to complete the strain step. Despite the relatively slow stretching rate, the first WAXD patterns after the strain step did not show any signs of crystallinity [37]. They suggested that crystallization proceeds on two time scales of approximately 0.1 and 4 s, respectively. To the author's knowledge, no step-like unloading experiments to study the disappearance of crystallites have been reported in literature. Le Cam noted the asymmetry of the relaxation behavior during loading and unloading, which he ascribed to different crystallization and melting kinetics [38]. A general interpretation of strain-induced crystallization on the molecular level, e. g. in analogy to the well-known nucleation and growth models for quiescent crystallization, or in connection with chain segment mobilities, is still lacking [26].

4.2.2 Dynamic Cyclic Experiments

Dynamic cyclic experiments were carried out on a selfmade electrodynamic tensile machine at a frequency of roughly 1 Hz. Figure 4.25a shows crystallinity over time for a cyclic experiment performed on unfilled NR. First, for $t < 0$ s, the sample was held at a constant strain of 365 % and a crystallinity of almost 6 % was observed. After starting the cyclic experiment (cycling between 0 % strain and 365 % strain) at $t = 0$ s, the crystallinity did not reach the same level as during the initial quasistatic part. Instead, the maximum crystallinity remained below 2 %. This can be ascribed to the finite crystallization kinetics described in the previous section. Because in each cycle the sample resides above the SIC onset strain for only a fraction of a second, time is too short to let the crystallinity reach the quasistatic level. Upon unloading, the sample becomes amorphous, as expected. No accumulation of crystallinity occured over time when the sample was extensively cycled.

Not only the maximum strain, but also the minimum strain (i.e. the R-ratio), determine crystallinity. This is shown in Fig. 4.25b, where the sample was cycled between 290 and 365 % strain—the same maximum strain as in the previous experiment. The crystallinity ranges around 4 %, which is still less than the quasistatic level of 6 %, but above the previous cyclic level of 2 %, when the sample was unloaded completely.

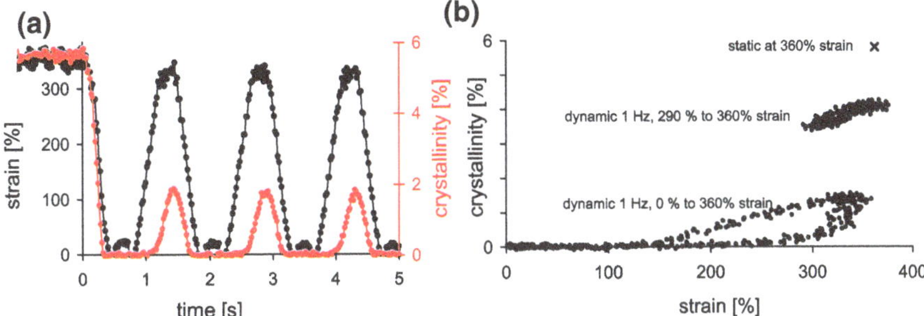

Fig. 4.25 **a** Crystallinity (*red*) and strain (*black*) over time, NR [#1] cycled between 0 and 365 % strain. **b** Crystallinity versus strain, NR [#1] cycled between 290 and 365 % strain. For comparison, the data from (**a**) is plotted

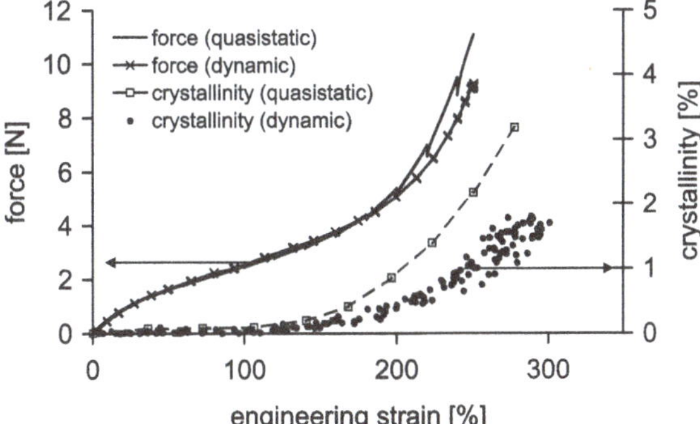

Fig. 4.26 Force and crystallinity versus engineering strain for filled NR [#11]. Dynamic and quasistatic experiments were performed as described in the text. For the dynamic experiments, for clarity, only the loading paths of 6 cycles are shown

Again, no accumulation of crystallinity over the number of cycles was observed. Hence these results disprove the hypothetical accumulation of crystallinity around a crack tip under non-relaxing loading conditions [20].

The SIC kinetics directly affect the mechanical behavior of NR. For instance, the disparity between the enormous resistance to static crack growth and the reduced resistance under dynamic conditions, is one of the consequences of the crystallization kinetics. In more general terms, the SIC kinetics add an unusual time-dependent component to the mechanical behavior of NR, which should also find its expression in constitutive models. In fact, when the strain amplitude of the dynamic tests exceeds the strain of SIC onset (approx. 170 % for filled NR [#11]), the lower crystallinity leads to a reduction in reinforcement and thus reduces the stress as compared to a quasistatic test, where sufficient time was given to reach crystallization equilibrium. This is exemplified in Fig. 4.26, where the same specimen was first cycled for 20 cycles ("demullinization", $\epsilon_{eng,min} = 0\,\%$, $\epsilon_{eng,max} = 250\,\%$), then the quasistatic force-displacement curve (with 30 s holding time in analogy to the beamline experiments) was recorded and finally the sample was cycled for another 20 cycles, 10 of which are shown. These mechanical experiments were performed ex situ. In order to avoid experimental inaccuracies, the engineering strain ϵ_{eng} from the readouts of the tensile machine are plotted. However, we confirmed that these are related to the optical strain data by a constant factor only. The amplitude of the in-situ dynamic experiments was slightly larger ($\epsilon_{eng,min} = 0\,\%$, $\epsilon_{eng,max} = 285\,\%$).

So far no direct measurements of crystallinity during cyclic dynamic loading have been published in literature. Stroboscopic techniques have been used to overcome the problem of limited X-ray beam flux, accumulating patterns over several mechanical cycles. The first study of SIC under dynamic load using this approach is due to Acken in 1932 [39]. He reported that the SIC onset under cyclic loading was shifted to greater strains compared to quasistatic loading. Since then, the stroboscopic technique was applied by several investigators. Kawai et al. [40, 41] reported that crystallization is in phase with stress and not with strain. Using different phase shifts between the stretching of the sample and the stroboscopic shutter, they were able to probe the evolution of crystallinity over the complete deformation cycle. In contrast to the results presented above, they reported that at frequencies below 2 Hz crystallization is independent of frequency. Recently, Candau et al. used the stroboscopic approach and deduced SIC incubation times of 50 ms [42], followed by an exponential increase of crystallinity over time according to a stretched exponential function with an exponent around 3. The inherent disadvantage of the stroboscopic technique is the accumulation over several cycles, such that no distinction can be made between structural evolution during any one cycle and a possible structural change over the course of several cycles.

4.3 Crystallinity Around a Crack Tip Under Dynamic Load

The combination of time-resolved and spatially-resolved techniques has become possible due to the high photon rate of the MiNaXS microbeam [43]. As described in Sect. 3.2.7, a sophisticated data processing is required to extract the structural information from huge amounts of scattering data. However, this kind of experiment most closely resembles the mechanical conditions the rubber is subjected to in tear fatigue experiments, justifying the experimental effort.

4.3.1 Comparison Between Static and Dynamic Loading

When exposing the sample (NR with 40 phr N234 [#25]) to a harmonic cyclic strain of 1 Hz with a minimum strain of 0 %, the maximum external strain without crack growth is 70 %. For comparison, under quasistatic conditions, the same sample can be stretched to more than 300 % until rupture occurs. The maximum crystallinity during cyclic loading is 1 % and the crystalline zone extends only about 100 μm into the sample. The maximum crystallinity is about half of the static level at the same external strain (Fig. 4.27). The complete series of crystallinity maps during cyclic

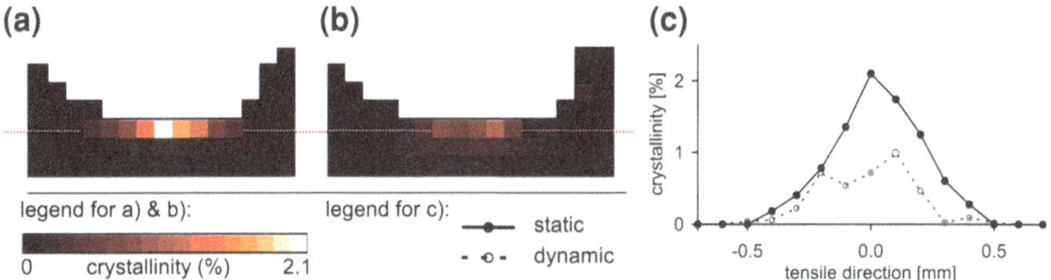

Fig. 4.27 **a** and **b** Crystallinity maps of NR with 40 phr N234 [#31] for an external strain of 70 %: **a** static, **b** dynamic. Both scans were performed on the same sample. *Each pixel* represents 100 × 100 μm. **c** Crystallinity versus horizontal position obtained from a slice through the crack tip along the profiles marked in (**a**) and (**b**)

loading is shown in Fig. 6.2. Rublon et al. performed similar experiments at the Soleil synchrotron [44]. Due to limited beam flux and detector readout times, the maximum in-situ mechanical frequency during pattern acquisition was 0.1 Hz, which is less than under typical fatigue conditions. Also, no comparison of crystallinity between static and dynamic loading was shown.

4.3.2 Influence of R-Ratio

The R-ratio is here defined as the ratio of the minimum and maximum strain under cyclic load. The same sample as in the previous section was subjected to a non-relaxing load ($R = 0.43$) with a minimum strain of 30 %, while the maximum strain was the same 70 % as in the fully relaxing experiment. In analogy to simple tensile experiments, one expects the crystallinity to be higher in the non-relaxing case. This was indeed confirmed. The total crystallinity plotted in Fig. 4.28 was obtained as the sum of the crystallinity values over all pixels of the crystallinity maps. The higher crystallinity in the non-relaxing case leads to a better crack growth resistance. Due to the stress concentration at the crack tip, the crystalline zone continues to exist even at the minimum strain of 30 %. Moreover, the maximum total crystallinity and also the maximum extension of the crystalline zone is not in phase with the external strain. Instead, the maximum is reached with a certain phase lag. This can be ascribed to the crystallization kinetics, especially in the areas a few 100 μm away from the crack tip. The highest single crystallinity value at the crack tip is observed in phase with the external strain.

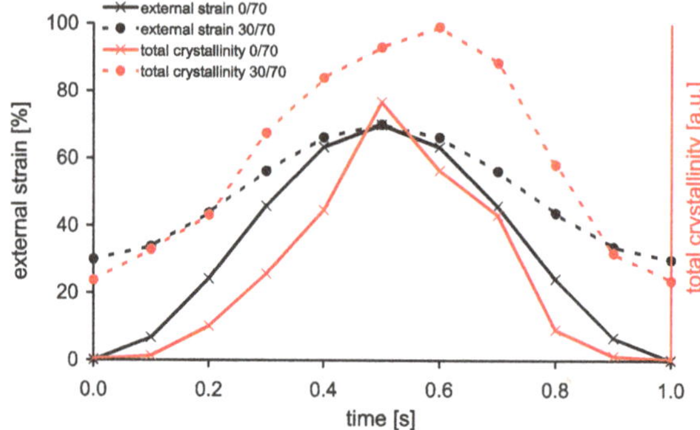

Fig. 4.28 External strain and total crystallinity over cycle time for a fully relaxing case (0/70) and a non-relaxing case (30/70) in NR with 40 phr N234 [#31]

4.4 Filler Orientation

As outlined in Sect. 3.2.9, spherical aluminum oxide and rod-like aragonite particles served as model fillers to study the orientation of fillers under strain and under fatigue loading. Due to the complexity of the carbon black structure and the simultaneous appearance of anisotropic cavities, a quantitative extraction of filler orientation from SAXS data of carbon black filled rubbers is till now impossible.

The USAXS patterns of NR with 10 phr Al_2O_3 are shown in Fig. 4.29a, b. As expected, the pattern remains isotropic under strain due to the spherical geometry of the filler. The same conclusion can be drawn from the WAXD patterns, where the Al_2O_3 rings have a constant intensity over the azimuthal angle. In contrast to this, the high aspect ratio aragonite filler orientates under strain. This is reflected in the anisotropy of the USAXS patterns (Fig. 4.29c–e). The iso-intensity lines turn from isotropic circular to ellipsoidal and finally, at large strains, assume a lozenge shape. The orientation is fully reversible when the load is removed.

The orientations of the patterns were quantitatively evaluated based on Eqs. 3.6–3.9. The Maier-Saupe orientational distribution parameter m as a function of strain is an input parameter to the USAXS models and was obtained from the WAXD patterns (Fig. 4.30). Under deformation, the aragonite diffraction rings turn into arcs, and increasingly narrow down at certain azimuthal angles, designating an orientation of the particles along the tensile direction. The orientation of the (111) plane is analyzed in Fig. 4.31a. Due to the position of the (111) plane in the crystal lattice, the diffraction intensity concentrates at an azimuthal angle $\alpha = 53°$ (with respect to the tensile direction), when the long axis of the cylindrical particles is aligned with the tensile direction. As shown in Fig. 4.31b, the orientational distribution can be well fit by a Maier-Saupe function, which increases almost linearly over strain. The scattering patterns obtained from Eq. 3.9 are depicted in Fig. 4.32. The qualitative

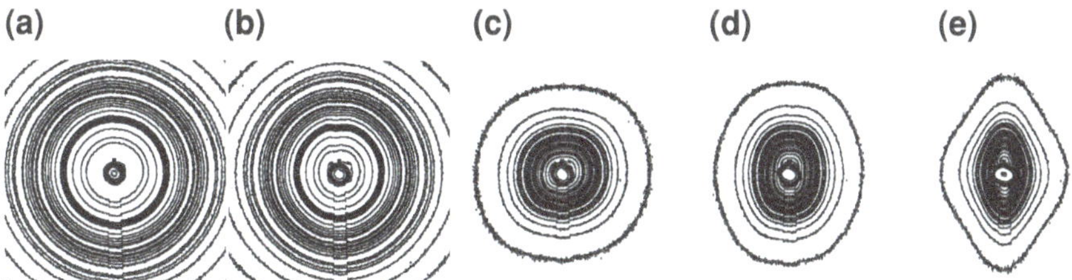

Fig. 4.29 USAXS patterns of NR with 10 phr Al_2O_3 [#32]: **a** undeformed, **b** 500 % strain; NR with 27 phr aragonite [#22]: **c** undeformed, **d** 150 % strain, **e** 500 % strain. The maximum scattering vector q is $0.12\,\text{nm}^{-1}$. Stretching direction is horizontal

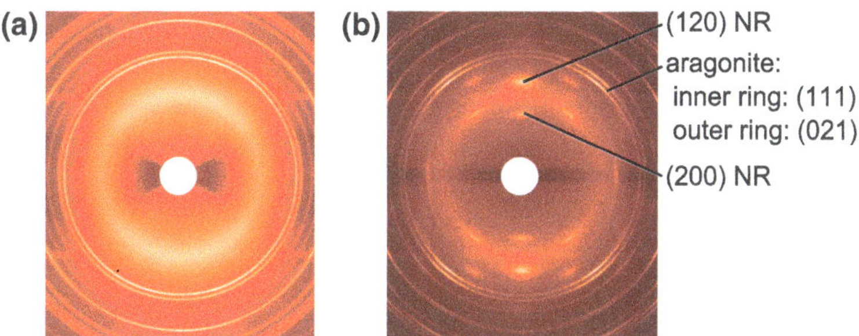

Fig. 4.30 WAXD patterns of NR with 27 phr aragonite [#22]. **a** Without deformation, **b** at 550 % strain. Stretching direction is *horizontal*

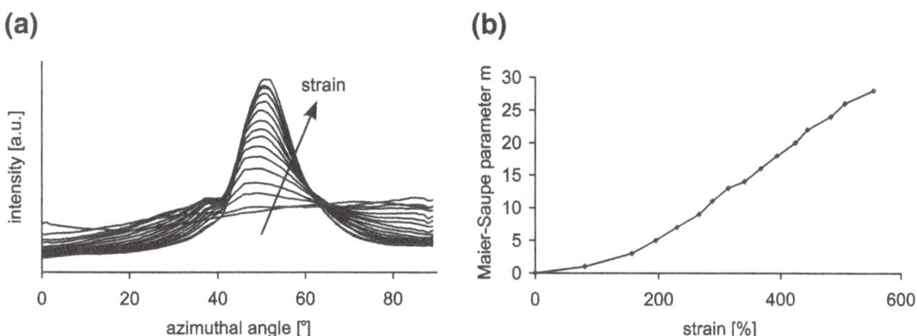

Fig. 4.31 **a** Intensity of the (111) peak of aragonite over the azimuthal angle α in NR filled with 27 phr aragonite [#22]. The strain increases from 0 to 550 %. The kink at 40° is an artifact due to a blind area at the boundary between two panels of the detector. **b** Maier-Saupe orientational distribution parameter derived from (**a**) with Eq. 3.6

agreement with the experimental patterns (Fig. 4.29) is obvious, validating the model assumptions.

To the author's knowledge, similar experiments with model compounds have so far not been reported in literature. Few studies dealt with the filler orientation in industrially relevant rubber composites filled with carbon black or silica. Ehrburger-

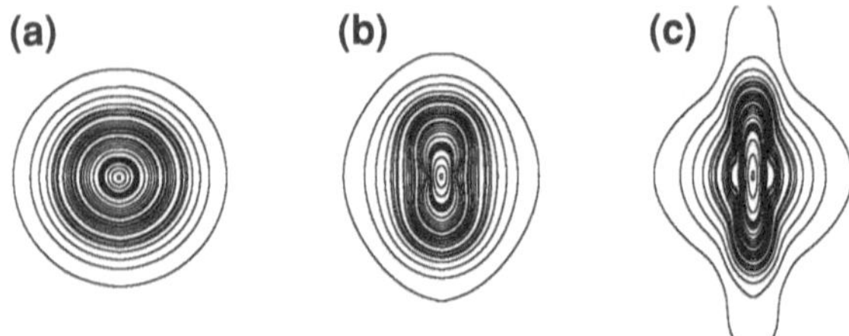

Fig. 4.32 Simulated scattering patterns acc. to Eq. 3.9 for rubber filled with cylindrical particles: **a** no deformation, **b** 150 % strain, **c** 500 % strain. The maximum scattering vector q is $0.12\,\mathrm{nm}^{-1}$. Stretching direction is *horizontal*

Dolle et al. reported anisotropic two-dimensional scattering patterns from stretched carbon black-filled rubber [45, 46]. However, the study was limited to small strains and no quantitative interpretation was given. Schneider studied silica-filled silicone and found an orientation of the aggregates along the stretching direction [47–49]. He developed a model based on self-affine fractals to interpret two-dimensional scattering data. However, no direct quantitative comparison between model and experimental data was given. The multi-scale deformation processes in rubbers filled with high-structure fillers are still under research.

4.5 Cavitation

Cavitation was studied by two independent methods. For a spatially resolved cavity map around a crack tip, USAXS was employed. However, this method relies on some approximations and a model. Optical volumetric studies served as model-free alternative method. A cross-check of all results by these independent methods is not always possible, since scattering experiments are limited to small and thin samples, while optical methods cannot yield local information.

4.5.1 Cavitation Studies by USAXS

Following the evaluation procedure specified in Sect. 3.2.10, we confirmed that in unfilled natural rubber, no increase in total scattering intensity could be detected in the USAXS regime and that this scattering regime is insensitive to crystallinity. The absence of cavities is in line with the literature, reporting constant volume for unfilled rubbers under deformation. Likewise, in moderately carbon black-filled rubbers (up to 30 phr) and in rubbers filled with non-reinforcing fillers (calcium carbonate, aluminum oxide) no cavitation was observed. Only in SBR and NR filled with at least 40 phr carbon black an increase in total scattering intensity was found (Fig. 4.33e). The increase occurs over a broad range of scattering vectors, hinting at a broad size distribution of cavities and rendering the evaluation of void fraction

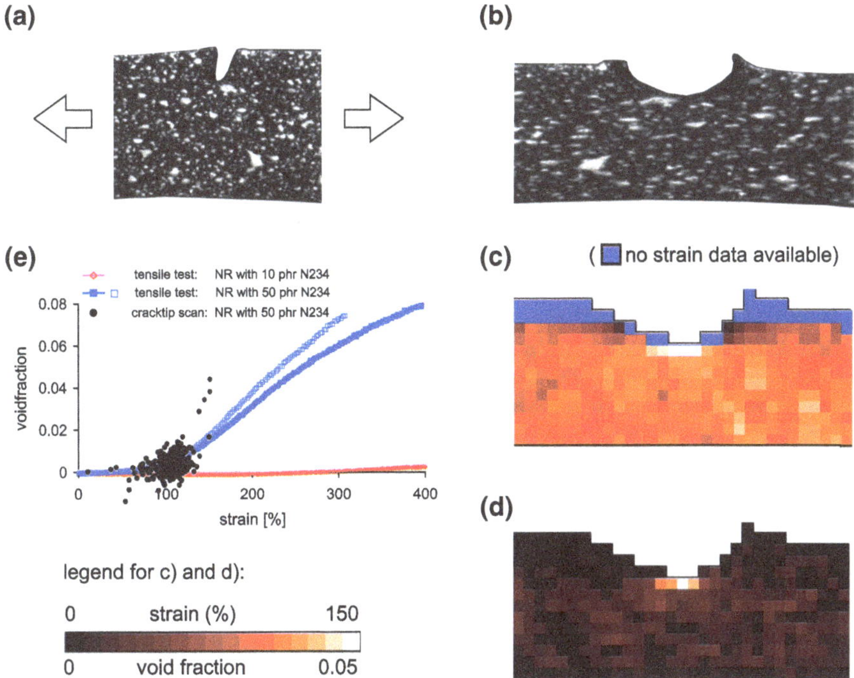

Fig. 4.33 Scanning USAXS performed on a strained tensile sample with a crack. The sample consists of NR filled with 50 phr N234 [#18]. **a** Photograph of the sample under very moderate strain. The speckle pattern serves for the strain field evaluation by digital image correlation. *Arrows* indicate the stretching direction. **b** Photograph of sample under high strain, the crack has propagated. **c** Strain field analysis (strain along the tensile direction). **d** Void fraction obtained by scanning USAXS. *Each pixel* represents $100 \times 100\,\mu m$. **e** Plot of void fraction over strain obtained from simple tensile experiments and from the scan [#9, #18]

relatively insensitive to the exact choice of the region of scattering angles over which the integration is performed to obtain the total scattering intensity.

Investigating the local structure around a crack tip by scanning USAXS with a spatial resolution of $100\,\mu m$ showed that cavitation occurs in a small zone directly in front of the crack tip (Fig. 4.33a–d). The maximum detected void fraction was around 0.05, but this value may vary with the step size of the scan. The magnitude of the void fraction is comparable to the values observed in a simple tensile sample under uniaxial strain (Fig. 4.33e). Similar values were also obtained for the same material under equibiaxial load.

Zhang et al. recently reported similar cavity volume fractions and similar effects of the filler content, studying carbon black-filled styrene-butadiene rubber under uniaxial tension [50].

4.5.2 Volumetric Changes Determined by Optical Methods

The global volume change was measured using a camera to follow the contours of a rotational symmetric specimen. This allows to calculate the specimen volume. Different specimen geometries with various aspect ratios (diameter vs. height) were tested. In low aspect ratio samples, the material is mainly subjected to uniaxial stresses. Transverse stresses can be considered negligible due to unrestricted lateral contraction. An example of this geometry is the dumbbell specimen (Fig. 3.3d). Unfilled natural rubber was tested and no change in volume was found within the accuracy of the method (1 %).

For larger aspect ratios, cylindrical specimens were glued between two metal plates. The disadvantage of this setup is, that the tests cannot be performed up to sample fracture because the glue fails prematurely. Figure 4.34 shows the results for two different materials and three different aspect ratios a. The geometries are listed in Table 6.1. While the volume changes are moderate for low aspect ratio samples, high aspect ratio samples show a large increase in volume, presumably caused by the hydrostatic stress state in the interior of the sample, promoting the formation of cavities, as outlined in literature [51, 52]. However, other factors, like internal decohesion at the metal-rubber interface, cannot be excluded as a cause for the increase in volume.

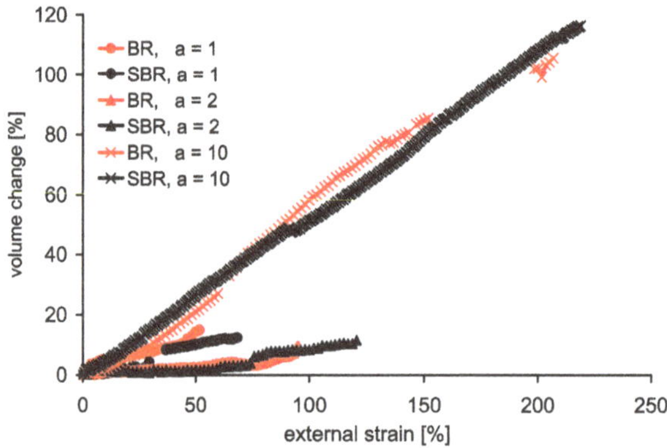

Fig. 4.34 Volume changes in cylindrical specimens under tension for SBR [#33] and BR [#34]

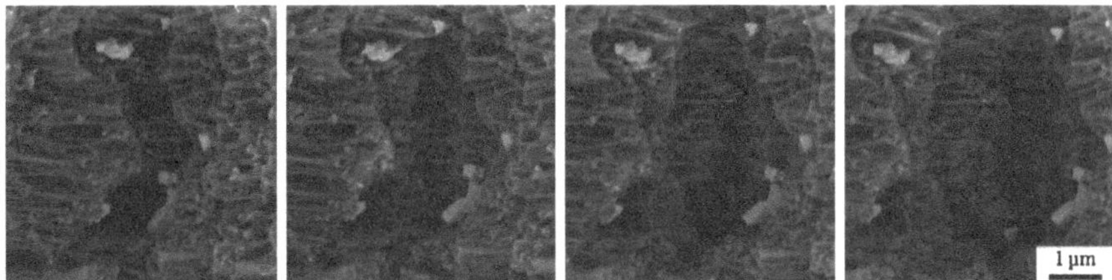

Fig. 4.35 Observation of crack opening in NR with 50 phr N234 [#35]. *Top view* (cf. Fig. 3.16b), facing the crack ground. Tensile direction is *horizontal*. The external displacement increases from *left* to *right*. Crack propagation reveals further layers of fibrillar material, similar to the top layer. The *scale bar* represents 1 μm. Potential sample damage by the electron beam cannot be excluded

4.6 Morphological Studies by Scanning Electron Microscopy

4.6.1 Real-Time Observation of Crack Growth

The growth of cracks in SENT samples was followed by in-situ tensile experiments in the SEM. Figure 4.35 shows a series of micrographs recorded with increasing strain. The crack opens, propagates and reveals underlying material, which appears to be similar to the material at the surface. The solid rubber bulk appears as porous fibrillar structure with highly oriented rubber strands. The precise nature of these strands is unclear. In NR, it suggests itself that they contain the highly oriented crystallites. However, strands are also observed in SBR (Fig. 6.5), which is in agreement with literature [53, 54]. The carbon black agglomerates are interspersed between the strands. These images were taken within a few minutes, and in order to propagate the crack, the strain was increased. However, crack growth could also be observed live, i.e. without increase of strain, over a few seconds, until a new stable state was reached. Unfortunately, these images could not be captured due to the low frame rate of high resolution images. The observed structure is similar to Fig. 4.35; underlying strands are revealed as soon as the top layer of strands breaks.

4.6.2 Structure of Carbon Black Around a Crack Tip

The structure of carbon black could be resolved down to the scale of individual primary particles (Fig. 4.36a). The adhesion between filler aggregates and rubber matrix apparently is strong, as evidenced by the tight bonding and the fact that they are well embedded in the matrix; no cavities and no free unbound carbon black are observed.[3] After fracture of the material, the orientated strands continue to exist. Most of them are capped with a carbon black aggregate (Fig. 4.36b), suggesting

[3] An attempt was made to quantify the orientation of the carbon black aggregates (Sect. A.3).

Fig. 4.36 SEM micrographs of a crack surface of NR with 60 phr N550 [#6] (*top view* (cf. Fig. 3.16b), tensile direction *horizontal*). **a** Notch groove under load. Individual carbon black aggregates are clearly visible. **b** Stress-free fracture surface after complete failure of the sample. Despite the removal of the load, some of the strands continue to exist. The *scale bars* represent 1 µm

that the ultimate failure took place at the interface between carbon black and rubber matrix.

To the author's knowledge, scanning electron microscopy studies revealing the structure down to carbon black primary particles have not previously been reported in literature [55–57]. Göritz performed in-situ tensile experiments on filled rubbers in a transmission electron microscope [54].

4.6.3 Strain Field on the Microscale

In order to resolve the strain field around a crack tip on the microscale, the notched sample was viewed from the side (Fig. 4.37). Characteristic features (like surface irregularities, filler particles etc.) were manually identified and corresponding strain values were evaluated semi-automatically.[4] The results show that the maximum strain is quite small (Fig. 4.38), and only a few tens of microns behind the crack tip, is less than 200 %, even when the sample is strained almost to fracture.

4.6.4 Adhesion of Zinc Oxide

The electron micrographs clearly show that the bonding between zinc oxide particles and rubber matrix is very weak (Fig. 4.39). After propagation of the crack, most particles lie on the surface, lacking any bonds to the rubber and also lacking

[4] A self-written software asks the user to click at the characteristic position in the image and then loads the next image (displaying the previous one for better orientation in a separate window). After the last image, the relative displacement vectors between certain characteristic features are computed.

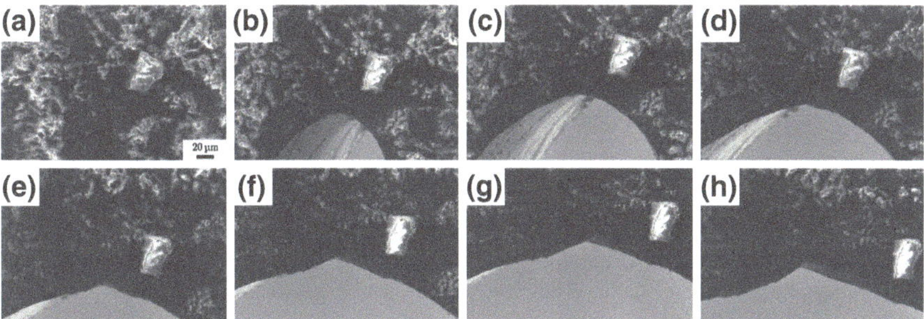

Fig. 4.37 SEM micrographs of NR with 90 phr N550 [#37] *side view* (cf. Fig. 3.16b). The *crack* starts to grow on a small length scale in **d**, and grows considerably in **f**. The *scale bar* represents 20 µm and applies to all images

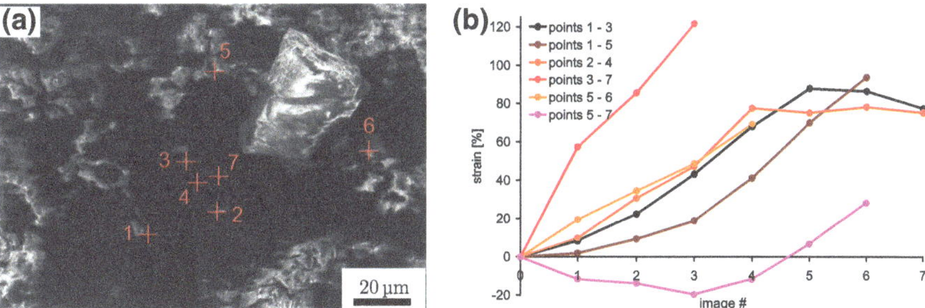

Fig. 4.38 **a** Assignment of characteristic points for strain evaluation. **b** Strain over image number for various combinations of points. The region in front of the crack tip (points 5–7) undergoes compression before crack growth, and elongation afterwards. The crack propagates between points 3 and 7, such that the strain can only be evaluated for four images

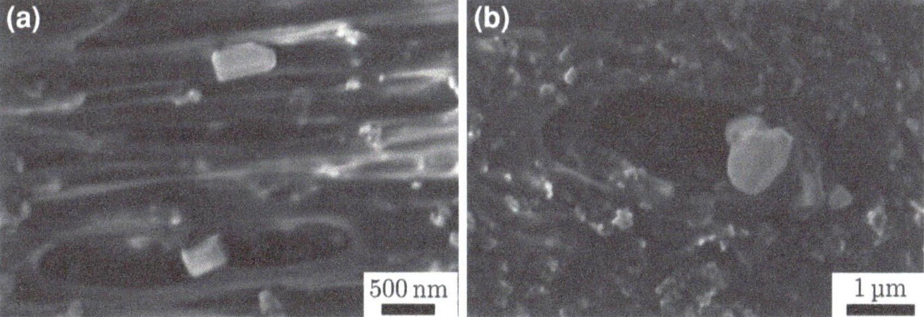

Fig. 4.39 SEM micrographs of NR with 50 phr N234 and 5 phr zinc oxide [#35]. (*Top view* of notch groove, tensile direction *horizontal*.) **a** Zinc oxide particles loosely lying on the surface, apparently not well bonded to the rubber. **b** Cavity caused by adhesive failure of the rubber-zinc oxide interface

any presence of bound rubber on their surface. These results are in line with work published by Le Cam [56].

References

1. B. Huneau, Strain-induced crystallization of natural rubber: a review of X-ray diffraction investigations. Rubber Chem. Technol. **84**(3), 425–452 (2011). doi:10.5254/1.3601131. http://link.aip.org/link/RCT/84/425/1
2. S. Poompradub et al., Mechanism of strain-induced crystallization in filled and unfilled natural rubber vulcanizates. J. Appl. Phys. **97**, 103529 (2005)
3. S. Trabelsi, P.A. Albouy, J. Rault, Effective local deformation in stretched filled rubber. Macromolecules **36**(24), 9093–9099 (2003)
4. J. Rault et al., Stress-induced crystallization and reinforcement in filled natural rubbers: 2H NMR study. Macromolecules **39**(24), 8356–8368 (2006)
5. S. Dupres et al., Local deformation in carbon black-filled polyisoprene rubbers studied by NMR and X-ray diffraction. Macromolecules **42**(7), 2634–2644 (2009)
6. S.D. Gehman, J.E. Field, X-ray structure of rubber-carbon black mixtures. Ind. Eng. Chem. **32**(10), 1401–1407 (1940)
7. S. Beurrot, B. Huneau, E. Verron, Strain-induced crystallization of natural rubber subjected to biaxial loading conditions as revealed by X-ray diffraction. In: Proceedings of the VII. European Conference on Constitutive Models for Rubber, pp. 23–28 (2011)
8. S.C. Nyburg, X-ray determination of crystallinity in deformed natural rubber. Br. J. Appl. Phys. **5**, 321–324 (1954)
9. J.M. Chenal et al., Parameters governing strain induced crystallization in filled natural rubber. Polymer **48**(23), 6893–6901 (2007)
10. S. Beurrot et al., In-situ synchrotron X-ray diffraction study of strain-induced crystallization of natural rubber during fatigue tests. In: Proceedings of the VII. European Conference on Constitutive Models for Rubber, pp. 353–358 (2011)
11. M. Klüppel, The role of disorder in filler reinforcement of elastomers on various length scales. Adv. Polym. Sci. **164**, 1–86 (2003)
12. L.E. Alexander, *X-ray Diffraction Methods in Polymer Science* (Wiley-Interscience, New York, 1969)
13. M. Tosaka et al., Orientation and crystallization of natural rubber network as revealed by WAXD using synchrotron radiation. Macromolecules **37**(9), 3299–3309 (2004)
14. M. Tosaka, Strain-induced crystallization of crosslinked natural rubber as revealed by X-ray diffraction using synchrotron radiation. Polym. J. **39**(12), 1207–1220 (2007)
15. L.B. Tunnicliffe, A.G. Thomas, J.J.C. Busfield, The effect of fillers on crosslinking, swelling and mechanical properties of peroxide-cured rubbers. In: Proceedings of the VIII. European Conference on Constitutive Models for Rubber. Balkema, pp. 563–568 (2013)
16. S. Trabelsi, P.A. Albouy, J. Rault, Stress-induced crystallization around a crack tip in natural rubber. Macromolecules **35**(27), 10054–10061 (2002)
17. S. Trabelsi, P.A. Albouy, J. Rault, Stress-induced crystallization properties of natural and synthetic cis-polyisoprene. Rubber Chem. Technol. **77**(2), 303–316 (2004)
18. D.J. Lee, J.A. Donovan, Microstructural changes in the crack tip region of carbon-black-filled natural rubber. Rubber Chem. Technol. **60**(5), 910–923 (1987)
19. J.B. Le Cam, E. Toussaint, The mechanism of fatigue crack growth in rubbers under severe loading: the effect of stress-induced crystallization. Macromolecules **43**(10), 4708–4714 (2010)
20. N. Saintier, G. Cailletaud, R. Piques, Cyclic loadings and crystallization of natural rubber: an explanation of fatigue crack propagation reinforcement under a positive loading ratio. Mater. Sci. Eng. A **528**(3), 1078–1086 (2011). ISSN: 0921–5093, doi:10.1016/j.msea.2010.09.079. http://www.sciencedirect.com/science/article/pii/S0921509310011251

21. M.M. Alam, H. Osaka, T. Asano, Biaxial orientation mechanism of drawn natural rubber. III. Determination of the orientation function from wide-angle X-ray diffraction pattern. J. Macromol. Sci. Part B: Phys. **47**(2), 317–337 (2008)
22. M.M. Alam, T. Asano, Biaxial orientation mechanism of drawn natural rubber. IV. The effect of aspect ratio. J. Macromol. Sci. B **47**(4), 754–764 (2008)
23. R. Oono, K. Miyasaka, K. Ishikawa, Crystallization kinetics of biaxially stretched natural rubber. J. Polym. Sci. Polym. Phys. Ed. **11**(8), 1477–1488 (1973)
24. K. Brüning et al., Kinetics of strain-induced crystallization in natural rubber studied by WAXD: dynamic and impact tensile experiments. Macromolecules **45**(19), 7914–7919 (2012)
25. L. Mandelkern, Crystallization kinetics of homopolymers: overall crystallization: a review. Biophys. Chemi. **112**(2–3), 109–116 (2004)
26. L. Mandelkern, *Crystallization of Polymers Volume 2: Kinetics and Mechanisms* (Cambridge University Press, Cambridge, 2004)
27. P.G. De Gennes, Coil-stretch transition of dilute flexible polymers under ultrahigh velocity gradients. J. Chem. Phys. **60**(12), 5030–5042 (1974)
28. R.H. Somani et al., Flow-induced shish-kebab precursor structures in entangled polymer melts. Polymer **46**(20), 8587–8623 (2005)
29. D.J. Dunning, P.J. Pennells, Effect of strain on rate of crystallization of natural rubber. Rubber Chem. Technol. **40**, 1381 (1967)
30. J.C. Mitchell, D.J. Meier, Rapid stress-induced crystallization in natural rubber. J. Polym. Sci. A-2 Polym. Phys. **6**(10), 1689–1703 (1968)
31. A.N. Gent, Crystallization and the relaxation of stress in stretched natural rubber vulcanizates. Trans. Faraday Soc. **50**, 521–533 (1954)
32. D. Luch, G.S.Y. Yeh, Strain-induced crystallization of natural rubber. III. Reexamination of axial-stress changes during oriented crystallization of natural rubber vulcanizates. J. Polym. Sci. Polym. Phys. Ed. **11**(3), 467–486 (1973)
33. H.G. Kim, L. Mandelkern, Crystallization kinetics of stretched natural rubber. J. Polym. Sci. A-2 Polym. Phys. **6**(1), 181–196 (1968)
34. G.S.Y. Yeh, Morphology of strain-induced crystallization of polymers. Part I. Rubber Chem. Technol. **50**, 863–873 (1977)
35. M. Tosaka et al., Crystallization of stretched network chains in cross-linked natural rubber. J. Appl. Phys. **101**, 084909 (2007)
36. M. Tosaka et al., Crystallization and stress relaxation in highly stretched samples of natural rubber and its synthetic analogue. Macromolecules **39**(15), 5100–5105 (2006)
37. M. Tosaka et al., Detection of fast and slow crystallization processes in instantaneouslystrained samples of cis-1,4-polyisoprene. Polymer **53**(3), 864–872 (2012)
38. J.B. Le Cam, E. Toussaint, Cyclic volume changes in rubber. Mech. Mater. **41**(7), 898–901 (2009)
39. M.F. Acken, W.E. Singer, W. Davey, X-ray study of rubber structure. Ind. Eng. Chem. **24**(1), 54–57 (1932)
40. H. Hiratsuka et al., Measurement of orientation crystallization rates of linear polymers by means of dynamic X-ray diffraction technique. II. Frequency dispersion of straininduced crystallization coefficient of natural rubber vulcanizates in subsonic range. J. Macromol. Sci. B **8**(1–2), 101–126 (1973). doi:10.1080/00222347308245796. eprint: http://www.tandfonline.com/doi/pdf/10.1080/00222347308245796. http://www.tandfonline.com/doi/abs/10/00222347308245796
41. H. Kawai, Dynamic X-ray diffraction technique for measuring rheo-optical properties of crystalline polymeric materials. Rheol. Acta **14**(1), 27–47 (1975). ISSN: 0035–4511. http://dx.doi.org/10.1007/BF01527209
42. N. Candau et al., Characteristic time of strain induced crystallization of crosslinked natural rubber. Polymer **53**(13), 2540–2543 (2012)
43. K. Brüning et al., Strain-induced crystallization around a crack tip in natural rubber under dynamic load. Polymer **54**(22), 6200–6205 (2013)

44. P. Rublon et al., In situ synchrotron wide-angle X-ray diffraction investigation of fatigue cracks in natural rubber. J. Synchrotron Radiat. **20**(1), 105–109 (2012)
45. F. Ehrburger-Dolle et al., Anisotropic ultra-small-angle X-ray scattering in carbon black filled polymers. Langmuir **17**(2), 329–334 (2001)
46. F. Ehrburger-Dolle, F. Bley, E. Geissler, F. Livet, I. Morfin, C. Rochas, Filler networks in elastomers. Macromol. Symp. **200**(1), 157–168 (2003). http://onlinelibrary.wiley.com/doi/10.1002/masy.200351016/
47. G.J. Schneider, Correlation of mass fractal dimension and asymmetry. J. Chem. Phys. **130**, 234912 (2009)
48. G.J. Schneider, D. Göritz, A novel model for the interpretation of small-angle scattering experiments of self-affine structures. J. Appl. Crystallogr. **43**(1), 12–16 (2009). ISSN: 0021–8898
49. G.J. Schneider, D. Göritz, Strain induced anisotropies in silica polydimethylsiloxane composites. J. Chem. Phys. **133**, 024903 (2010)
50. H. Zhang et al., Nanocavitation in carbon black filled Styrene-butadiene rubber under tension detected by real time small angle X-ray scattering. Macromolecules **45**(3), 1529–1543 (2012)
51. K. Legorju-Jago, C. Bathias, Fatigue initiation and propagation in natural and synthetic rubbers. Int. J. Fatigue **24**(2-4), 85–92 (2002)
52. N. Aït Hocine et al., Experimental and finite element investigation of void nucleation in rubber-like materials. Int. J. Solids Struct. **48**(9), 1248–1254 (2011)
53. G. Michler, *Electron Microscopy of Polymers* (Springer, Berlin, 2008)
54. D. Göritz, Properties of rubber elastic systems at large strains. Macromol. Mater. Eng. **202**(1), 309–329 (1992)
55. K. Agarwal, D.K. Setua, K. Sekhar, Scanning electron microscopy study on the influence of temperature on tear strength and failure mechanism of natural rubber vulcanizates. Polym. Test. **24**(6), 781–789 (2005)
56. J.B. Le Cam et al., Mechanism of fatigue crack growth in carbon black filled natural rubber. Macromolecules **37**(13), 5011–5017 (2004)
57. S. Beurrot, B. Huneau, E. Verron, In situ SEM study of fatigue crack growth mechanism in carbon black-filled natural rubber. J. Appl. Polym. Sci. **117**(3), 1260–1269 (2010)

Chapter 5
Conclusion and Outlook

Owing to the combination of novel experimental methods and automatized data processing routines, new insight was gained into the structural behavior of natural rubber under various mechanical loads. The employment of third generation synchrotron radiation opened the possibility to study strain-induced crystallization in natural rubber on previously inaccessibly short time scales. Tensile impact experiments, performed in analogy to pressure- or temperature jump experiments in chemical reaction kinetics analysis, proved the finite kinetics of strain-induced crystallization and a first phenomenological kinetic law was established [1]. The crystallization proceeds on a time scale which is longer than the mechanical time scale typically encountered in elastomer products in most applications, e.g. tires. This implies that the structure obtained under such dynamic loadings is far from equilibrium, which was directly confirmed by in-situ dynamic cyclic tests in the synchrotron. These tests closely resemble the conditions of tear fatigue tests. In fact, the mechanical behavior was shown to qualitatively correlate with the structural properties. For instance, the discrepancy between the enormous crack growth resistance of natural rubber under static loading and the significantly lower resistance under dynamic loading can directly be traced back to the kinetics of strain-induced crystallization. Also the faster crack growth under pulse load as compared to harmonic load is a direct consequence, as is the sensitivity to the R-ratio.

One of the most outstanding properties of natural rubber is its superior crack growth resistance due to the self-reinforcing ability of the crystallites, especially under static loading. WAXD crack tip scans under static and dynamic load with an unprecedented spatial resolution revealed the influence of the strain rate on the crystallinity around the crack tip, which is reflected in the strain rate dependency of the crack growth resistance.

Cavitation, often discussed as an initiating step in crack growth, was studied by two independent methods. For the first time, a USAXS scan quantitatively revealed cavitation processes around a crack tip. Secondly, a new method was established to study the volume change of elastomers under large strains by automatized evaluation

K. Brüning, *In-situ Structure Characterization of Elastomers during Deformation and Fracture*, Springer Theses, DOI: 10.1007/978-3-319-06907-4_5,

of camera images. It was shown that the formation of cavities becomes significant under loading conditions in which hydrostatic stresses prevail.

Because almost all technical rubber goods are filler reinforced, also the structural changes of the filler under strain is of great interest. While a conclusive model for the quantitative interpretation of anisotropic USAXS patterns of carbon black filled rubbers is still lacking, a new model could be established for geometrically simpler model fillers of spherical or rod-like shape [2]. It was shown that the filler orientation is reversible under fatigue loading.

In particular, this work contributed to substantial progress in the following areas:

- A novel tensile impact machine was designed to perform synchrotron WAXD experiments with an unprecedented time resolution of 7 ms, giving new insight into the very early stages of strain-induced crystallization in natural rubber.
- A new tensile machine was built for in-situ dynamic cyclic experiments at a frequency of 1 Hz, which allowed to observe the real-time evolution of crystallinity under cyclic loading for the first time.
- It was shown that the crystallinity under dynamic cyclic load increases with increasing R-ratio.
- A local map of the crystallinity distribution and orientation around a crack tip under static conditions was obtained from WAXD crack tip scans with an unprecedented spatial resolution.
- An elaborate method was developed to carry out a crack tip scan under tear fatigue conditions (1 Hz) for the first time, yielding a crystallinity map under real-life loading conditions.
- A model, based on first principles, was developed to interpret SAXS patterns of rubbers filled with anisometric or isometric particles in order to quantify the evolution of particle orientation with strain.
- A micro tensile machine for in situ experiments in a SEM allowed to observe the processes around a propagating crack in real time on length scales below 1 μm.
- A method was developed to measure the volumetric strain in rubbers under deformation by optical means.

These methodological advances allowed to establish new structure-property relationships, especially linking the time dependency of the mechanical and fracture behavior of rubbers with the strain-induced crystallization:

- The tensile impact experiments revealed that the nucleation in strain-induced crystallization is much faster than previously thought.
- A phenomenological law for the description of the crystallization kinetics at room temperature was established. Thus the critical strain rate can be estimated at which the time-dependency of strain-induced crystallization becomes important.
- The dynamic cyclic experiments clearly revealed that for dynamic loading conditions, as found in most rubber products, the hitherto performed quasistatic WAXD experiments are only of limited significance. Together with the static and dynamic crack tip scans, they instructively illuminate the underlying fundamentals behind the huge discrepancy between static and dynamic crack growth experiments, which is typical for strain-crystallizing elastomers.

- The underlying structural reasons for the R-ratio dependence of the tear fatigue resistance of strain-crystallizing elastomers were elucidated qualitatively.
- On a microscale, crack propagation can be considered a self-similar process, i.e. upon crack growth the typical fibrillar structure is already established in layers underneath the crack surface.

Besides the progress made, the findings also raised new questions. In particular, the following issues are waiting to be tackled by researchers:

- Can one interpret the experimentally found crystallization kinetics in terms of a molecular model, connecting the macroscopic degree of crystallinity with molecular dynamics, such as segmental relaxation times or bond angle rotations?
- Is it possible to set up a physically motivated constitutive model, considering the time-dependence of the reinforcing strain-induced crystallites as strain-rate and history depending crosslinks or filler particles?
- Can one extend the proposed model for anisometric filler particles to real fillers, like carbon black or silica, to gain a better understanding of the reinforcement mechanisms and of the anisotropy of filled rubbers under strain?
- Can the formation of cavities in a notched sample in front of a crack tip, or in a bulk sample under hydrostatic stress, be confirmed by absolute methods, such as computed tomography?
- How does biaxial strain influence the crystallization? How can one obtain a crystallinity map of a non-fiber symmetric biaxially loaded sample with reasonable experimental effort?

In conclusion, the application of a series of novel and sophisticated experimental techniques gave new insight into structure-property relationships of filled rubbers under deformation. This hopefully contributes to a more fundamental understanding of the mechanical behavior of reinforced rubbers. The long-term vision is to replace the currently empirical tasks of designing compound formulations and rubber components by an approach from first principles.

References

1. K. Brüning et al., Kinetics of strain-induced crystallization in natural rubber studied by WAXD: dynamic and impact tensile experiments. Macromolecules **45**(19), 7914–7919 (2012)
2. K. Brüning, K. Schneider, G. Heinrich, Deformation and orientation in filled rubbers on the nano- and microscale studied by X-ray scattering. J. Polymer Sci. B Polymer Phys. **50**(24), 1728–1732 (2012)

Appendix

A.1 Conductivity of Samples for Scanning Electron Microscopy

See Fig. A.1.

A.2 Crystallinity Maps Under Dynamic Loading

The crystallinity maps at 10 equidistant steps (in time) during a cyclic sinusoidal loading (minimum strain 0 %, maximum strain 70 %) of NR with 40 phr N234 [#31] were extracted from the series of diffraction patterns to create a video to illustrate the evolution of crystallinity in real time. Shown below (Fig. A.2) is the series of images that the video is composed of.

A.3 Orientation of Carbon Black Aggregates from SEM

An attempt was made to quantify the orientation of the carbon black aggregates as seen by SEM. A PV-Wave code was written to distinguish carbon black aggregates from surrounding matrix and automatically decompose the micrograph into a two-phase black and white image (Fig. A.3). Then, the chord length distribution of the carbon black phase was computed along the tensile direction and along the transversal direction. An orientation was inferred from the fact that along the tensile direction larger chord lengths prevail, whereas along the transversal direction, smaller chord lengths dominate (Fig. A.4).

The code performs an edge enhancement and identifies the carbon black by its higher intensity. A frequency filtering in the Fourier space tries to filter out strands.

K. Brüning, *In-situ Structure Characterization of Elastomers during Deformation and Fracture*, Springer Theses, DOI: 10.1007/978-3-319-06907-4,

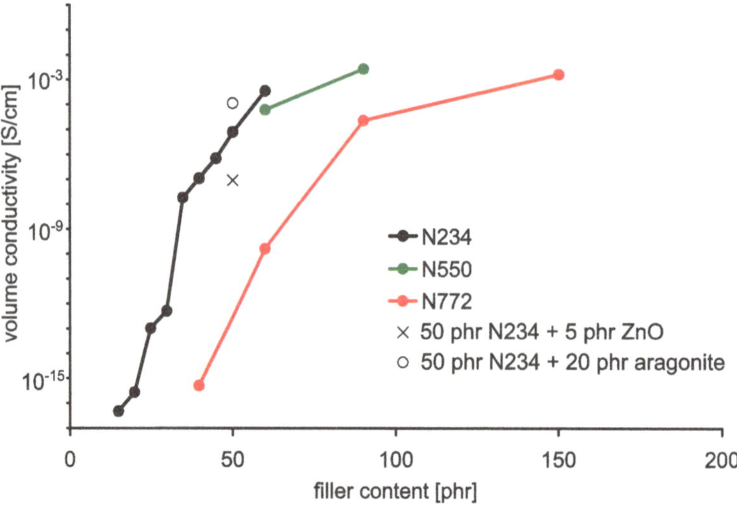

Fig. A.1 Volume conductivity of NR compounds with various filler contents and filler grades [#10, #11, #12, #13, #14, #15, #16, #17, #19, #36, #37, #27, #28, #29, #30, #35]

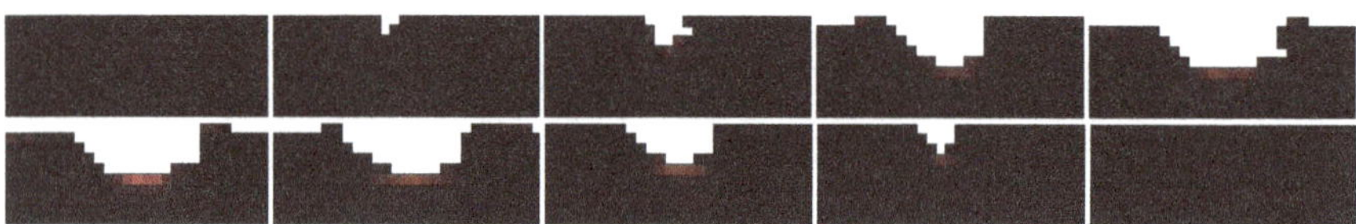

Fig. A.2 Crystallinity maps for a complete cycle (minimum strain 0 %, maximum strain 70 %) of NR with 40 phr N234 [#31]. The color scale is the same as in Fig. 4.27

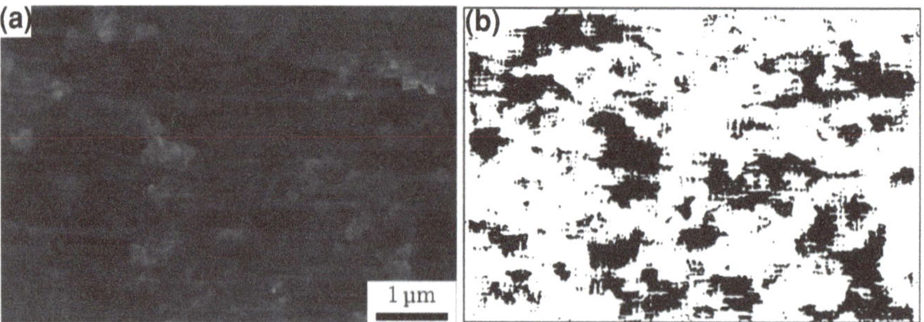

Fig. A.3 **a** SEM micrograph of a crack surface of NR with 60 phr N550 [#6] (*top view* (cf. Fig. 3.16b), tensile direction horizontal). **b** Conversion of **a** into a two-phase image. The scale bar represents 1 μm

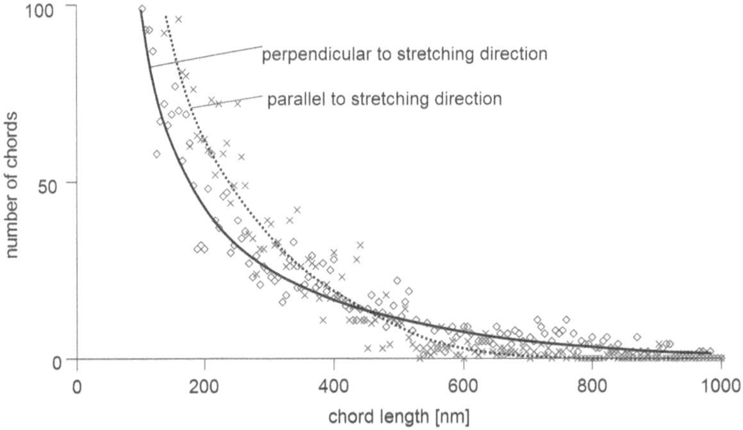

Fig. A.4 Chord length distribution from Fig. A.3b. One could infer an orientation from the shift of the chord length distribution to larger chords along the tensile direction

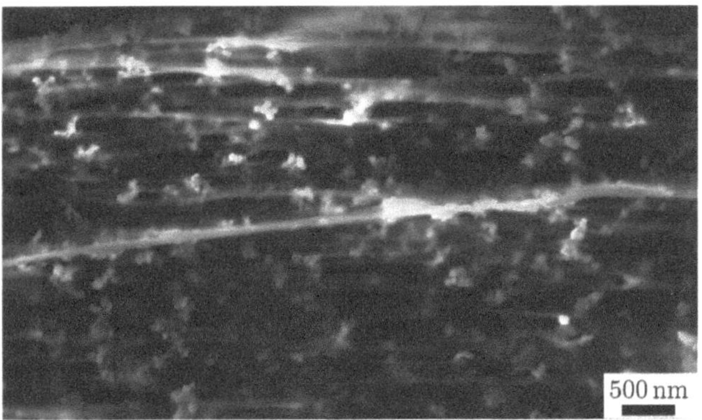

Fig. A.5 SEM micrograph of a crack surface (stretched in situ) of SBR with 40 phr N339 [#39] (*top view*, tensile direction horizontal). Fibrillar strands are equally visible as in stretched NR, despite the absence of crystallites

However, the code is not able to unequivocally distinguish between strands and carbon black. Due to the strong anisotropy of the strands, small errors in this step have huge effects on the chord length distribution. Finally, this method was discarded.

A.4 SEM of Carbon Black Filled SBR

See Fig. A.5.

Table A.1 Specimen geometries of cylindrical samples for optical volumetric measurements

Aspect ratio a	Diameter (mm)	Height (mm)
1	18	18
2	18	9
10	20	2

Fig. A.6 Photograph of pale crepe natural rubber as supplied

A.5 Specimen Geometries of Cylindrical Samples for Optical Volumetric Measurements

See Table A.1.

6 List of Materials

ID	Matrix	Filler	Zinc oxide	DCP[a]	Sulfur	Stearic acid	Others	Vulcanization[b]
#1	NR[c]			2 phr				19 min at 160 °C
#2[d]	NR		3 phr		1.7 phr	1 phr	2.5 phr CBS[e], 1.5 phr IPPD[f]	7.8 min at 160 °C
#3[d]	NR	20 phr N234	3 phr		1.7 phr	1 phr	2.5 phr CBS[e], 1.5 phr IPPD[f]	4.7 min at 160 °C
#4[d]	NR	40 phr N234	3 phr		1.7 phr	1 phr	2.5 phr CBS[e], 1.5 phr IPPD[f]	3.7 min at 160 °C
#5[d]	NR	60 phr N234	3 phr		1.7 phr	1 phr	2.5 phr CBS[e], 1.5 phr IPPD[f]	2.5 min at 160 °C
#6[d]	NR	60 phr N550	3 phr		1.7 phr	1 phr	2.5 phr CBS[e], 3.4 phr IPPD[f]	2.5 min at 160 °C
#7[d]	NR	40 phr N339	n. a.		n. a.	n. a.	n. a.	3.7 min at 160 °C
#8[d]	NR	40 phr N339g[h]	n. a.		n. a.	n. a.	n. a.	6.1 min at 170 °C
#9	NR[c]	10 phr N234		2 phr[i]				18 min at 160 °C
#10	NR[c]	15 phr N234[j]		2 phr				18 min at 160 °C
#11	NR[c]	20 phr N234[j]		2 phr				18 min at 160 °C
#12	NR[c]	25 phr N234[j]		2 phr[i]				18 min at 160 °C
#13	NR[c]	30 phr N234[j]		2 phr[i]				17 min at 160 °C

ID	Matrix	Filler	Zinc oxide	DCP[a]	Sulfur	Stearic acid	Others	Vulcanization[b]
#14	NR[c]	35 phr N234[j]		2 phr[i]				17 min at 160 °C
#15	NR[c]	40 phr N234[j]		2 phr[i]				17 min at 160 °C
#16	NR[c]	45 phr N234[j]		2 phr[i]				17 min at 160 °C
#17	NR[c]	50 phr N234[j]		2 phr[i]				17 min at 160 °C
#18	NR[c]	50 phr N234[j]		1 phr[i]				15 min at 160 °C
#19	NR[c]	60 phr N234[j]		2 phr[i]				16 min at 160 °C
#20	NR[c]	1 phr aragonite[k]		2 phr[i]		3 phr		20 min at 160 °C
#21	NR[c]	10 phr aragonite[k]		2 phr[i]				22 min at 160 °C

[a]Dicumyl peroxide
[b]The listed times are t_{90} at the given temperatures. Depending on the geometry, additional time is allowed (Sect. 3.1.2)
[c]Pale crepe NR, supplied by Weber und Schaer, Hamburg, Fig. A.6
[d]Supplied by Deutsches Institut für Kautschuktechnologie (DIK) as green compound
[e]N-cyclohexyl-2-benzothiazole-sulfenamide
[f]N-isopropyl-N′-phenyl-p-phenylene diamine
[g]Supplied by Columbian Carbon
[h]Graphitized N339
[i]Supplied by Acros Organics
[j]supplied by Evonik
[k]Custom grade, supplied by Schaefer Kalk

ID	Matrix	Filler	Zinc oxide	DCP	Sulfur	Stearic acid	Others	Vulcanization
#22	NR[a]	27 phr aragonite[b]		2 phr[c]		3 phr[d]		17 min at 160 °C
#23	NR[a]	54 phr aragonite[b]		2 phr[c]		3 phr[d]		15 min at 160 °C
#24	NR[a]	81 phr aragonite[b]		2 phr[c]		3 phr[d]		13 min at 160 °C
#25	NR[a]	60 phr N234[e]	3 phr[c]		1 phr[c]	1 phr[d]	1.5 phr CBS[f,g]	3.1 min at 160 °C
#26	IR[h]			2 phr[c]				19 min at 160 °C
#27	NR[a]	40 phr N772[e]		1.5 phr[c]		5 phr[d]		15 min at 160 °C
#28	NR[a]	60 phr N772[e]		1.5 phr[c]		5 phr[d]		20 min at 160 °C
#29	NR[a]	90 phr N772[e]		1.5 phr[c]		15 phr[d]		18 min at 160 °C
#30	NR[a]	150 phr N772[e]		1.5 phr[c]		15 phr[d]		12 min at 160 °C
#31	NR[a]	40 phr N234[e]	3 phr[c]		1 phr[c]	1 phr[d]	1.5 phr CBS[f,g]	3.4 min at 160 °C
#32	NR[a]	10 phr Al2O3[i]		2 phr[c]				19 min at 160 °C
#33	SBR 1502	10 phr CB		1.5 phr			3 phr ZnO2, 1 phr AO[j], 1 phr ACC[k]	n. a.
#34	BR 150L			1.5 phr			3 phr ZnO2, 1 phr AO[j], 1 phr ACC[k]	n. a.
#35	NR[a]	50 phr N234[e]	5 phr[c]	1.5 phr[c]		5 phr[d]		13 min at 160 °C

ID	Matrix	Filler	Zinc oxide	DCP	Sulfur	Stearic acid	Others	Vulcanization
#36	NR[a]	60 phr N550[e]		1.5 phr[c]		5 phr[d]		16 min at 160 °C
#37	NR[a]	90 phr N550[e]		1.5 phr[c]		15 phr[d]		18 min at 160 °C
#38	NR[a]	50 phr N234[e,l]		1.5 phr[c]		5 phr[d]		7.5 min at 160 °C
#39[m]	SBR	40 phr N339		n. a.		n. a.		5.9 min at 170 °C
#40[m]	SBR			3 phr	1.7 phr	1 phr	2.5 phr CBS[f], 1.5 phr IPPD[n]	21 min at 160 °C
#41[m]	SBR	20 phr N234[g]	3 phr		1.7 phr	1 phr	2.5 phr CBS[f], 1.5 phr IPPD[n]	9.7 min at 160 °C
#42[m]	SBR	40 phr N234[g]	3 phr		1.7 phr	1 phr	2.5 phr CBS[f], 1.5 phr IPPD[n]	7.0 min at 160 °C
#43[m]	SBR	60 phr N234[g]	3 phr		1.7 phr	1 phr	2.5 phr CBS[f], 1.5 phr IPPD[n]	5.1 min at 160 °C

[a]Pale crepe NR, supplied by Weber und Schaer, Hamburg, Fig. A.6
[b]Custom grade, supplied by Schaefer Kalk
[c]Supplied by Acros Organics
[d]Supplied by Fisher Scientific
[e]Supplied by Evonik
[f]N-cyclohexyl-2-benzothiazole-sulfenamide
[g]Supplied by Lanxess
[h]Natsyn 2200, supplied by Goodyear Chemical
[i]Nanophase Nanodur
[j]Antioxidant, N-tert-butyl-2-benzothiazole-sulfenamide
[k]Accelerator, 1,3-diphenylguanidine
[l]Second filler: 20 phr aragonite[b]
[m]Supplied by Deutsches Institut für Kautschuktechnologie (DIK) as green compound
[n]N-isopropyl-N′-phenyl-p-phenylene diamine

Curriculum Vitae

	Karsten Brüning born June 5, 1985; Oldenburg
Education	
2013	Ph.D. in Engineering (Dr.-Ing.) at Technische Universität Dresden/Leibniz-Institut für Polymerforschung Dresden (Prof. G. Heinrich, Dr. K. Schneider): In-situ Structure Characterization of Elastomers during Deformation and Fracture
2009	Diploma thesis at Leibniz-Institut für Polymerforschung Dresden (Prof. G. Heinrich, Dr. R. Vogel) in cooperation with the Institute of Medical and Polymer Engineering, TUM (Prof. E. Wintermantel): Fabrication and Characterization of Melt-Spun Fibers made from Electron-Irradiated PEEK
2007–2008	Studies of Materials Science and Engineering at Georgia Institute of Technology, Degree: M. Sc.
2004–2009	Studies of Chemical Engineering at Technische Universität München, Degree: Dipl.-Ing.
2004	Abitur (Gymnasium Brake)
Work Experience	
2013–current	Scientist at Leibniz-Institut für Polymerforschung Dresden, Institute for Polymer Materials (Prof. G. Heinrich), Department Mechanics and Structure (Dr. K. Schneider)
2009–2013	Ph.D. studies at Leibniz-Institut für Polymerforschung Dresden, Institute for Polymer Materials (Prof. G. Heinrich), Department Mechanics and Structure (Dr. K. Schneider)
Honors and Awards	
2014	Tire Technology International Young Scientist Award
2013	Contribution to DESY Photon Science Highlights Report 2013
2010	First Price Poster Award, International Conference on X-Ray Investigations in Polymer Sciences, Wroclaw
2007–2008	World Student Fund Scholarship

K. Brüning, *In-situ Structure Characterization of Elastomers during Deformation and Fracture*, Springer Theses, DOI: 10.1007/978-3-319-06907-4,

List of Publications

Journal articles

1. K. Brüning, K. Schneider, S.V. Roth, G. Heinrich: Kinetics of strain-induced crystallization in natural rubber studied by WAXD: Dynamic and impact tensile experiments. Macromolecules 45 (2012), 7914-7919
2. K. Brüning, K. Schneider, K. Heinrich: Deformation and orientation in filled rubbers on the nano- and microscale studied by X-ray scattering. Journal of Polymer Science Part B: Polymer Physics 50 (2012), 1728-1732
3. K. Brüning, K. Schneider, S.V. Roth, G. Heinrich: Strain-induced crystallization around a crack tip in natural rubber under dynamic load, Polymer 54 (2013), 6200-6205
4. X. Zhang, K. Schneider, G. Liu, J. Chen, K. Brüning, D. Wang, M. Stamm: Deformation-mediated superstructures and cavitation of poly(L-lactide): In-situ small-angle X-ray scattering study. Polymer 53 (2012), 648-656
5. X. Zhang, K. Schneider, G. Liu, J. Chen, K. Brüning, D. Wang, M. Stamm: Structure variation of tensile-deformed amorphous poly(L-lactic acid): Effects of deformation rate and strain, Polymer 52 (2011), 4141-4149

Book chapters

1. K. Brüning, K. Schneider, G. Heinrich: In-situ structural characterization of rubber during deformation and fracture; in: M. Klüppel, W. Grellmann, G. Heinrich, K. Schneider, M. Kaliske, T. Vilgis (Editors): Fracture Mechanics and Statistical Mechanics of Reinforced Elastomeric Blends (Lecture Notes in Applied and Computational Mechanics), 43-80, Springer 2013

Conference Proceedings

1. K. Brüning, K. Schneider, G. Heinrich: Short-time behavior of strain-induced crystallization in natural rubber. Proceedings of the VIII. European Conference on Constitutive Models for Rubber (ECCMR 8), 2013, 457-460, CRC Press, London

K. Brüning, *In-situ Structure Characterization of Elastomers during Deformation and Fracture*, Springer Theses, DOI: 10.1007/978-3-319-06907-4,

2. K. Brüning, K. Schneider, G. Heinrich: Filler orientation in filled elastomers under deformation. Proceedings of the 15th Small-Angle Scattering Conference (SAS 2012), Sydney, Australia
3. K. Brüning, K. Schneider, G. Heinrich: Strain-induced crystallization in natural rubber under dynamic mechanical load. Proceedings of the 10. Kautschuk Herbst Kolloquium (KHK 2012), Hannover
4. K. Brüning, K. Schneider, G. Heinrich: The kinetics of strain-induced crystallization in natural rubber. Proceedings of the 15th International Conference Polymeric Materials (P2012), Halle, 2012
5. K. Brüning, K. Schneider, G. Heinrich: Deformation mechanisms in carbon blackfilled rubbers on the nano- and microscale. Proceedings of Deformation, Yield and Fracture of Polymers (DYFP) 2012, Kerkrade, Netherlands
6. K. Brüning, K. Schneider: Effects of strain field and strain history on the natural rubber matrix. Proceedings of the VII. European Conference on Constitutive Modelsfor Rubber (ECCMR 7), 2011, 29-32, CRC Press, London
7. K. Brüning, K. Schneider, G. Heinrich: Filler-related phenomena in elastomers under mechanical load studied by X-Ray scattering. Proceedings of the Eurofillers Conference 2011, Dresden
8. K. Brüning, K. Schneider: Filled Rubbers under Mechanical Load studied by X-Ray Scattering. Tagungsband Deformation und Bruchverhalten von Kunststoffen 2011, Merseburg
9. K. Brüning, K. Schneider: Morphology evolution in filled rubbers under cyclic and fatigue loading using USAXS, SAXS and WAXS. Proceedings of the VIII. International Conference on X-Ray Investigations of Polymer Structures (XIPS) 2010, Wroclaw, Poland
10. K. Brüning, K. Schneider: Strain-induced crystallization and void formation in on-line stretched natural rubber studied by synchrotron SAXS and WAXS. Proceedings of the 14th International Conference Polymeric Materials (P2010), Halle

Oral Presentations

1. K. Brüning, K. Schneider, G. Heinrich: Short-time behavior of strain-induced crystallization in natural rubber. VIII. European Conference on Constitutive Models for Rubber (ECCMR 8), 25.06. - 28.06.2013, San Sebastian, Spain
2. K. Brüning, K. Schneider, G. Heinrich: Strain-induced crystallization in natural rubber: Structure-property relations under realistic loading conditions. International Conference for Rubber and Rubberlike Materials 2013 (ICRRM), 07.03. - 09.03.2013, Kharagpur, India
3. K. Brüning, K. Schneider, G. Heinrich: The kinetics of strain-induced crystallization in natural rubber: Time-resolved in-situ dynamic and impact tensileWAXD experiments. IPF Kolloquium Programmbereich 3, 20.09.2012
4. K. Brüning, K. Schneider, G. Heinrich: The kinetics of strain-induced crystallization in natural rubber. 15th International Conference Polymeric Materials (P2012), Halle, 12.09. - 14.09.2012

5. K. Brüning, K. Schneider, G. Heinrich: Deformation mechanisms in carbon blackfilled rubbers on the nano- and microscale. Deformation, Yield and Fracture of Polymers (DYFP), Kerkrade, Netherlands, 01.04. - 05.04.2012
6. K. Brüning, K. Schneider: Effects of strain field and strain history on the natural rubber matrix. VII. European Conference on Constitutive Models for Rubber (ECCMR), Dublin, Ireland, 20.09. - 23.09.2011
7. K. Brüning, K. Schneider, G. Heinrich: Deformation Mechanisms in Carbon Black Filled Rubber on the Nanoscale. Eurofillers Conference 2011, Dresden, 21.08. - 25.08.2011
8. K. Brüning, K. Schneider, G. Heinrich: Filled Rubbers under Mechanical Load studied by X-Ray Scattering. Deformation und Bruchverhalten von Kunststoffen, Merseburg, 30.06. - 01.07.2011

Poster Presentations

1. K. Brüning, K. Schneider, S. V. Roth, G. Heinrich: Self-reinforcement of rubber: Crystallization under dynamic strain. Hasylab User Meeting 2013, Hamburg, 24.01. - 25.01.2013
2. K. Brüning, K. Schneider, G. Heinrich: Filler orientation in filled elastomers under deformation. 15th Small-Angle Scattering Conference, Sydney, Australia, 18.11. - 23.11.2012
3. K. Brüning, K. Schneider, G. Heinrich: Strain-induced crystallization in natural rubber under dynamic mechanical load. 10. Kautschuk Herbst Kolloquium (KHK 2012), Hannover, 07.11. - 09.11.2012
4. K. Brüning, K. Schneider: WAXD study on the kinetics of strain-induced crystallization in natural rubber, Hasylab User Meeting 2012, Hamburg, 26.01.-27.01.2012
5. K. Brüning, K. Schneider: Morphology evolution in filled rubbers under cyclic and fatigue loading using USAXS, SAXS and WAXS. VIII. International Conference on X-Ray Investigations of Polymer Structures (XIPS) 2010, Wroclaw, Poland, 08.12. - 10.12.2010, First Poster Price Award
6. K. Brüning, K. Schneider: Simultaneous SAXS and WAXS to study strain-induced crystallization and void formation in on-line stretched natural rubber. 14th International Conference Polymeric Materials, 15.09. - 17.09.2010, Halle

Reports

1. K. Brüning, K. Schneider, G. Heinrich: Transient phenomena in strain-induced crystallization in natural rubber. Jahresbericht 2012 des Leibniz-Instituts für Polymerforschung Dresden e.V.
2. K. Brüning, K. Schneider, S.V. Roth, G. Heinrich: Self-reinforcement of rubber - crystallization under dynamic strain. DESY Photon Science Highlights Report 2012
3. K. Brüning, K. Schneider, G. Heinrich: Dynamic crack tip scans of strain-crystallizing natural rubber. DESY Hasylab Yearbook 2012
4. K. Brüning, K. Schneider, G. Heinrich: Strain-induced crystallization in natural rubber under dynamic mechanical load. DESY Hasylab Yearbook 2011

5. K. Brüning, K. Schneider, G. Heinrich: Morphological changes in filled rubbers under mechanical load. DESY Hasylab Yearbook 2010

Other References

1. K-Zeitung, Ausgabe 9, 03.05.2013, p. 13: Stabilitätstest von Naturkautschuk
2. IPF intern, Heft 62 (Januar 2013), p. 6: Einem Geheimnis der Natur auf der Spur: Stabilität von Naturkautschuk
3. DESY website, Hasylab News, 23.09.2012: WAXD measurements at PETRA III: The kinetics of strain-induced crystallization in natural rubber

Index

K. Brüning, *In-situ Structure Characterization of Elastomers during Deformation and Fracture*, Springer Theses, DOI: 10.1007/978-3-319-06907-4,